V-Bombers

With the dawning of the atomic age in August 1945 it was obvious to military strategists that any future bomber to deliver a nuclear weapon would have to be jet powered and of advanced aerodynamic design. In the late 1940s and early 1950s the threat of a nuclear war spurred nations to develop fleets of long-range four-engine jet bombers capable of carrying such a nuclear payload. Towards the end of the Second World War Sir Frederick Handley Page, A. V. Roe and Vickers, who had provided the aircraft for RAF Bomber Command in over five years of war, had already begun thinking in terms of large jet-propelled bombers. In 1946 discussions of their proposals were made with the Air Staff and they resulted in specification B35/46 being finalised by the end of the year. The Royal Air Force Staff requirements for such an aircraft were covered in specification B.35/46 and this was issued to the aircraft industry on New Year's Day 1947. The specification for a medium-range bomber with a range of 3,500 nautical miles, which was capable of carrying a 10,000lb bomb to its target was issued to seven aircraft constructors and tenders requested. Of those submitted the two selected to go ahead were from Handley Page and A.V. Roe. At Shorts of Belfast meanwhile, work had begun on the Sperrin S.A.4, a shoulder-wing monoplane design with four 6,000lb s.t. Avon R.A.2 engines mounted on the wings. Only two Sperrins were built, the first flying on 10 August 1951. Though the Sperrin enjoyed the distinction of becoming the first of the British four-jet bombers, a month later a new specification was issued to cover the production of the Vickers Type 660, which would become the Nation's first four-jet bomber to enter production as the Valiant. Specifications for a more advanced design than the Sperrin were issued to elicit proposals from Handley Page, A. V. Roe and Vickers, who, following their wartime experience with the Halifax, Lancaster and Wellington respectively, were proven specialists in bomber development. While the Handley Page and Avro jet bomber projects were of an advanced nature their design was such that a vast amount of research was needed to ensure their ultimate success. At the time, the requirements of B35/46 were very advanced and far-reaching and Handley Page expected that it would take until 1952 before a prototype could be flown.

Such were the advanced requirements of the B.35/46 specification that the Air Staff had decided on an 'insurance' design for an aeroplane of more conventional nature which might guarantee earlier availability for RAF service. Vickers' interim V-Bomber design complied with a slightly less demanding (B.9/48) specification and on 2 February 1949 the company received an order for two prototypes: the Type 660 with four 6,500lb.s.t. Avon R.A.3s and the Type 667 powered by four Sapphires. At Vickers George R. Edwards and his design team evolved a bomber with wings which embodied compound sweepback in two degrees on the leading edge, the engines being concealed entirely within the centre-section and mounted adjacent to each other close to the fuselage. The tailplane was raised well above turbulence from the wings and the jet efflux by installing it towards midway on the fin. The Valiant prototype made its maiden flight on 18 May 1951 but it was destroyed during a test flight over Hampshire on 12 January 1952. (A second

Handley Page Victor K.1 tanker and Avro Vulcan refuelling. *MoD*

Handley Page Victor B.1 taxiing at Changi, Singapore in 1965. In the background are Avro Shackletons of No.205 Squadron and Hawker Siddeley Argosys of No.215 Squadron. *Jerry Cullum*

Avro Vulcan B.2s XH658 (nearest) and XH657 of No.101 Squadron over Ballistic Missile Early Warning System (BMEWS), Fylingdales, Yorkshire. *MoD*

was built and flew for the first time on 11 April that same year). On 5 September 1951 a new requirement was issued to start production of the Valiant B.Mk.1. The first of five pre-production aircraft flew at Brooklands on 22 December 1953.

At Avro meanwhile, the design team led by Roy Chadwick had originally decided on a conventional layout with swept wings but by progressively shortening the fuselage and then removing the tailplane, a delta-wing planform emerged. This was a bold step by the designers because little was known about delta aerodynamics at this time and the concept of the 'flying triangle' of these proportions was a daunting prospect. By March 1947 Chadwick had made a firm decision to go ahead with this configuration. Research into delta-wing aircraft had also been conducted by the Royal Aircraft Establishment (RAE) at Farnborough and the combined results of this programme and that of the Avro team brought about a number of changes in the design. The wing became thinner, with the engines buried inside and their intakes built into the wing-

roots and wing-tip fins were deleted in favour of a large central fin. The Avro 698 design was submitted to the Air Ministry in May 1947 by which time the triangle design had grown a nose with large engine intakes at each side. That summer the Ministry of Supply received the tender for the Avro 698 but the Avro team suffered a grievous loss with the death of Roy Chadwick who was killed in the crash of the Tudor 2 prototype at Woodford on 23 August. It was feared that Chadwick's death might cause the Ministry to lose confidence in the new delta but the Ministry had accepted Avros proposals on 23 July. Assistant Chief Designer S. D. Davies who had survived the accident now took over design leadership of the delta programme. The order covering two Avro 698 prototypes was received in January 1948 and work commenced immediately. To investigate the practically unknown characteristics of the delta design it was decided that a series of one-third scale research-aircraft would be built and these were designated the Avro 707s with the design work commencing in May of that year.

The first Avro 707 was a fairly simple aeroplane with an unusual engine air intake built into the top of the fuselage behind the cockpit. To speed construction the aircraft utilised the cockpit and nose-wheel leg of a Gloster Meteor and the main undercarriage of an Avro Athena. This aircraft was flown for the first time on 4 September 1949 by Avro test pilot S. E. 'Red' Else, who two days later flew it to Farnborough where it appeared in the famous Air Show, but on 30 September it crashed near Blackbushe and Else was killed. In all, five small Avro 707 series deltas of varying designs were constructed, each powered by a 3,500lb s.t. Rolls-Royce Derwent engine and they provided valuable research to the Avro 698 programme before the bomber prototype flew in 1952. The last of the Avro deltas flew in 1953, by which time the prototype 698 (VX770) had made its maiden flight (on 30 August 1952 in the capable hands of Avro's chief test pilot Wing Commander R. J. 'Roly' Falk). The chosen power plant, the 9,750lb s.t. Bristol Olympus B. E. 10, was not yet ready when the Avro 698 was approaching completion and four Rolls-Royce 6,500lb s.t. Avon R.A.3s powered the prototype at the time of its first flight. In January 1953 the name Vulcan was chosen for the Avro 698, the first bomber in the world to adopt the delta wing planform. This completed the famous RAF V-Bomber trio of Valiant, Victor and Vulcan. Later, VX770 was powered by Armstrong Siddeley Sapphires before the second prototype (VX777) made its initial flight on 3 September 1953 with four Bristol Olympus 101 engines (the aircraft crashed at RAF Syreston during the 14 September 1958 Battle of Britain display with the loss of all four crew). The 6,500lb thrust Rolls-Royce Avons were fitted as standard to all the production Vul-

cans, the first of which made its maiden flight on 4 February 1955. Despite its enormous size the Vulcan could demonstrate almost fighter-like manoeuvrability at very low altitudes. At the Farnborough Air Show in September 1955 test pilot Roly Falk caused a sensation by slow-rolling the aircraft during its flyby.

At Handley Page meanwhile, Reginald S. Stafford designed the HP 80 to meet Specification B.35/46 issued to meet the requirements of the Valiant and Vulcan. The company adopted the same approach as Avro had done by using a small flying scale model to test the design concept. In 1948 the HP 88 single-seat research aircraft was built by Blackburn as the Y.B.2 with a one-quarter-scale version of the proposed shape of the wings and the tail unit of the Victor. The H.P.88 flew on 21 June 1951 but it crashed at Stansted later that same year, killing its pilot D. J. P. 'Duggie' Broomfield. Stafford's HP.80 design included wings incorporating three successively decreasing degrees of sweepback from roots to tips in what became known popularly as the crescent wing. It was anticipated that the resulting reduction of drag would enable a high critical-Mach number to be reached at a constant value across the span and at the same time the crescent wing of high aspect ratio was thought to offer excellent range and extreme operational altitude. It was hoped that the first HP80 would be flying by the time of the SBAC (Society of British Aircraft Constructors) Show in September 1952. But the HP 80 was not ready to fly until a few days before Christmas and when it was ready the weather prevented the first flight from taking place until Christmas Eve. After Christmas, test flying continued at Boscombe Down until mid-February 1953, by which time the

Avro Vulcan B.2 XM606. *MoD*

HP80 had been given the name 'Victor' and it then returned to Radlett where a runway extension had been completed. Flight testing during the remainder of 1953 and the first half of 1954 demonstrated generally good handling qualities up to an altitude of 50,000 feet and an IMN (Indicated Mach Number) of 0.925. 'The only real problem', recalled test pilot John Allam OBE FRAeS, 'was with insufficient directional stability resulting in an undesirable Dutch-roll, which the pilot could control but this prevented him from ever relaxing'. Disaster struck on 14 July 1954 when the first prototype [WB771] was undertaking position-error calibration tests at Cranfield. During what was to be the final run at low altitude along the Cranfield runway the tail plane separated from the top of the fin and the aircraft crashed killing the crew. At this time the second prototype was almost ready to fly but its first flight was delayed while the cause of the accident was investigated and modifications made to its tail plane/fin attachments. It eventually flew on 11 September 1954 and made its second flight that day flying in the 1954 SBAC Show.

The Valiant was the first of the V Class four-jet bombers to enter RAF service, replacing the Canberras and Lincolns of 3 Group Bomber Command in 1955 and all 104 Valiant production aircraft had been delivered by the end of 1957 when they equipped nine squadrons. Powered by four 10,000lb thrust Rolls-Royce Avons, they were able to carry a 10,000lb nuclear bomb on a 3,500-mile sortie, cruising at 0.76 mach at 35–40,000 feet. In June 1955 No.232 Operational Conversion Unit (OCU) formed at Gaydon to train Valiant aircrews and initially, ground crews. The first course became the nucleus of 138 Squadron, which transferred from Gaydon to Wittering to become the RAF's first V-bomber squadron

on 6 July 1955. On 18 November 1955 No.543 Squadron, which moved from Gaydon to Wyton became the second Valiant squadron to be formed. On 21 January 1956 No.214 Squadron was re-formed at Marham in Norfolk and it took delivery of its first Vickers Valiant B(PR).1 to become the third V-bomber unit. Valiants were followed into service by eleven long-range strategic reconnaissance versions, the B(PR).1 and fourteen B(PR)K.1s and 45 BK.2s. Both the latter marks were capable of operating as a bomber or as a flight-refuelling tanker.

On 1 February 1956 the first production Victor B.Mk.1 flew and joined the prototype in the development programme. **John Allam** recalls that: 'It flew with the shorter fin and the definitive nose, which was 40 inches longer than the prototype noses. Other production Victors started to follow and by the end of August 1956 the seventh production aircraft had flown. All the Victors joined the flight test and development programme undertaking various tasks including high altitude handling and performance, flight flutter, position error calibration, auto-stabilizer, auto leading edge flap trials, auto-pilot, bombing, and a multitude of engineering and electrical trials. Aircraft had been flown to over 52,000 feet and at IMNs up to 0.98. In June 1957 the first production aircraft exceeded Mach 1 in a shallow dive at about 40,000 feet, producing witnessed supersonic bangs in the Watford area. It was the first large four-engine jet aircraft to achieve supersonic flight.'

On 1 April 1956 No.207 Squadron was re-formed as the fourth Valiant squadron and 214 Squadron re-equipped with the B.(PR)K.Mk.1. This multi-purpose version of the Valiant was capable of operating as a bomber or reconnaissance or as a flight refuelling tanker. On 23 April three Valiants took

Vickers Valiant B.1s of No.18 Squadron at RAF Finningley where they shared a joint ECM role with Canberra B.2s before the squadron disbanded on 31 March 1963. *Vickers*

part in a flypast at Marham in the presence of Premier Mr. Nikita Krushchev and other distinguished visitors from the Soviet Union that included Marshal Bulganin and A. N. Tupolev, the great Russian bomber design-bureau chief. On 1 May No.90 Squadron disbanded at Marham and in mid-July 35 Squadron moved to Upwood. In June a Valiant from each of Nos 214 and 543 Squadrons were flown to Idris, 15 miles south of Tripoli in Libya to take part in Operation Thunderhead to test NATO defences in the Mediterranean theatre and southern Europe. On 1 July Nos 214 and 207 Squadrons were joined at Marham by No.148 Squadron flying Valiants. On 26 July 1956 international tensions were heightened again when President Gamal Abdel Nasser of Egypt announced that his government intended to nationalise the Suez Canal. Britain and France were determined to reverse the decision by military means and Operation Musketeer, a joint Anglo-French undertaking, was put into action to destroy the Egyptian Air Force. All front line stations were immediately brought to operational readiness as aircraft were prepared for the preliminary deployment to Malta and Cyprus.

On 24 September four Valiants (three from No.214 Squadron and one from No.207 Squadron) left Marham for Luqa airfield in Malta as part of Operation Albert. Group Captain L. Hodges, the Marham Station Commander, acted as Force Commander. The main force of what was to be known as the Valiant Wing arrived on the island on 26 October. It comprised twenty-four Valiants made up of four from No.214 Squadron commanded by Wing Commander Leonard Trent VC DFC, six of No.207 Squadron commanded by Wing Commander D. Haig, six of No.148 Squadron commanded by Wing Commander W. Burnett while another eight were from No.138 Squadron at Wittering. Though designed to drop thermonuclear bombs the Valiants were fitted with multi-carriers and six light-series bomb racks to carry conventional free-fall bombs, the biggest being the 1,000 pounder. Twenty-one of these weapons could be carried. Chances of accurate delivery were poor however as many of the Valiants were bereft of their full complement of navigational and radar-operated bombing equipment. Instead they were forced to rely on the World War Two system of target marking while ex-Lincoln bomber sighting heads were fixed in temporary fittings. (As it turned out, those Valiants that carried navigation and bombing systems (NBS) most went unserviceable in the air). Bomber Command was ill prepared to undertake a Musketeer-type operation. The Command was geared to a 'radar' war in Western Europe and was not constituted nor organised for major overseas operations. The majority of the Valiant force had neither Navigation Bombing Systems nor visual bombsights and was not cleared for HE stores. The Canberra aircraft forming the bulk of the force deployed were equipped only with Gee-H as a blind bombing device and it was not possible to position ground-based beacons to give coverage for this equipment over Egypt. It was considered that it would be prudent for the Valiants and Hal Far- and Cyprus-based Canberras to carry out night bombing attacks on Abu Sueir, Kabrit, Almaza, Fayid and Cairo West, the main Il-28 base, and this necessitated a reversion to the marking technique successfully used in WWII. The plan was to crater the runways, followed up at dawn by ground attack strikes to destroy the aircraft. It was considered that the destruction of the Egyptian Airforce (EAF) would be achieved in two days. Little opposition was expected.

Flying Officer R. A. C. Ellicott of 214 Squadron recalls: 'The looks and expressions of surprise can only be imagined when, within two hours of landing at Luqa, all crews gathered in the Bomber Wing Operations briefing room for the first operational briefing and the curtains were drawn aside to reveal Egyptian airfields as the targets. Targets in Phase I were the Egyptian airfields operating Russian-built Il-28 bombers and MiG-15 fighters. The aiming points were the runway intersections and crews were briefed to avoid the camp areas. Further instructions were given that bombs were not to be jettisoned "live" in case Egyptian casualties were caused. At dusk on 30 October operations commenced.'

On the night of 30 October/1 November the Valiants and Canberras set off to bomb eight of the nine airfields in the Canal Zone and four more in the Nile Delta with 500lb and 1,000lb bombs. Five Valiants from No.148 Squadron and one from No.214 Squadron plus four Canberras from No.109 Squadron and three from No.12 Squadron attacked Almaza airfield near Cairo at 19.00Z hours. Intelligence reports stated that there were ten

Handley Page Victor B.1 XA918 at Farnborough. *John Hosford*

The single prototype Vickers Type 673 B.Mk.2 WJ954. *Vickers*

Blue Steel stand-off bomb at RAF Scampton. *Ashley Annis*

Vampires, ten MiG-15s, ten Il-28 bombers, nine Meteors and thirty-one twin-engine transports on the airfield. Canberras from No.139 Squadron operating from Cyprus did the visual marking. The first Red Target Indicator (TI) markers from a PFF Canberra went down over Heliopolis and the flares were dropped on the western hardstands 1,000 yards from the nominated aiming point. A second marker dropped TIs closer to the aiming point and called the bombers to drop on the most eastern set of TIs. Bombing was scattered, the Valiants attacking from 42,000 feet with free-fall bombs. XD814 from No.148 Squadron was the first V-bomber in action. Because of the poor marking, the 50 per cent error circle was 1,550 yards and only one runway was hit, suffering superficial damage. It was a similar story at Kabrit and Abu Sueir where, although accuracy was better, damage was light. Little opposition was encountered; there was light AA fire in the target area but it was sporadic and well below the attacking aircraft. The Valiant piloted by Squadron Leader Trevor Ware was intercepted by an Egyptian Meteor NF.13 but its pilot could not hold on to the jet bomber, which climbed out of his range and the Meteor flew away below the bomber.

The following night attacks were made on the airfields at Cairo West, Fayid and Kasfareet. Damage was again light and at Cairo West, the Ilyushins had already flown to Luxor. On the night of 2/3 November seven Valiants attacked Huckstep Barracks, which was bombed again two nights later together with the coastal batteries on El Agami Island.

During the attacks on Huckstep the markers were well placed and bombing was reasonably concentrated but at El Agami most of the bombers failed to drop because the TIs were extinguished, probably having fallen into the sea. Although Anti-Aircraft Armament (AAA) defended many of the targets, this was usually light and almost always well below the aircraft. By 3 November the EAF had virtually ceased to exist. About 260 of its aircraft had been destroyed, including most of the 120 MiGs and 50 Il-28s that were in service when the conflict began. The Anglo-French airborne assault began on 5 November and the seaborne assault went ahead on the morning of the 6th when all immediate objectives had been taken. But mounting pressure at the UN instigated by the USA led to a ceasefire on 8 November and the RAF and French crews stood down. The total Bomber Command effort during the Suez campaign amounted to forty-nine Valiant and 278 Canberra sorties. Post-conflict analysis determined that the destruction of the EAF was largely due to attacks by the fighter-bombers. On 17 November all the Valiants returned to the United Kingdom.

On 11 October 1956 a Valiant of No.49 Squadron carried the first British operational atomic bomb to be dropped from an aircraft, which was released over Maralinga, Southern Australia. On 15 May 1957 a Valiant of the same Squadron dropped the first British Hydrogen bomb during Operation Grapple on Christmas Island in the Pacific.

The first Vulcan B.Mk.1 squadron to be formed was No.83 at Waddington on 21 May 1957, with four aircraft and five crews from the first No.230 Operational Conversion Unit (OCU) course. They were followed by Nos 101 and 617 'Dambusters' at Finningley and Scampton respectively. From the time of their arrival into service at the OCU all the Vulcans were painted silver, but in April 1957 the first wearing the all-new white colour scheme arrived and the earlier aircraft were gradually repainted. Altogether, forty-five Mk.1s were built and ultimately equipped eight squadrons. The B.1 could reach a height of about 55,000 feet and fly at a maximum speed of about 620mph. The crescent-wing Victor was the third and final

V-bomber to be built for the RAF. At this time it was generally accepted that current high-speed, high-altitude jet bombers could out-fly interceptors of the day. However, its development was so protracted that by the time it entered operational service with the RAF with No.10 Squadron at Cottesmore in April 1958, it was no longer immune to fighters and missiles.

The first Victor was delivered to the RAF Victor OCU at Gaydon at the end of November 1957. Additional aircraft followed rapidly and Victor crew training started in earnest. The next Victors to be delivered to service were three specially-equipped radar reconnaissance aircraft to No.543 Squadron at Wyton to take over the work that had previously been carried out by Valiants. The first operational Victor bomber squadron (No.10) started to take delivery of its aircraft in April 1958 and the second squadron (No.15) in September 1958. Other squadrons followed in turn and the Victor assumed its role as part of the nuclear deterrent force. The Victors also started to fly around the world in connection with their various duty and training requirements and some of the sorties involved produced fast flights. One aircraft flying from Goose Bay, Labrador, to Marham, took 4 hours 1 minute from take-off to landing giving an average speed of 618 mph. Its coast-to-coast time, a distance of 2,020 miles, was 3 hours 8 minutes giving an average speed of 644 mph. Another Victor flew from Farnborough to Luqa, Malta, in exactly 2 hours, a distance of 1,310 miles, improving

on the then existing record for the same leg which was held by a Supermarine Scimitar.

In February 1958 successful air-to-air refuelling trials were carried out using Valiant tanker and receiver aircraft from No.214 Squadron. In 1952 the Air Staff had decided to adopt the in-flight refuelling system for the V-bomber force. Some Valiants would be equipped as tankers with removable refuelling equipment and all Valiant, Vulcan and Victor aircraft would be capable of receiving fuel. Design studies began in 1953 and early in 1954 a Valiant B.Mk.2 engaged in formating trials with a Canberra tanker. Trials with a production Valiant B.Mk.1 were carried out during 1956, and Valiant-to-Valiant trials began in 1957. By April 1958 the technique had been cleared for Service use. Forty-five Valiant B.K.1s equipped for bomber or tanker duties were built and this variant was used by No.214 Squadron for flight refuelling trials in 1959. During the trials, which ended in May 1960, several long-distance refuelled flights were made and the compatibility was demonstrated with the other V-bombers. No.90 Squadron converted to the tanker role in August 1961 but both units retained their bomber commitments until April 1962, when they officially became tanker squadrons. Operational squadrons began converting to the receiver role during 1960. Victor squadrons began tanker-receiving conversion during 1962.

In October 1958 the Valiants were invited to take part in the annual SAC Bombing

The flight of the second prototype Vulcan had come just in time for this aircraft to take part in the 1953 SBAC display at Farnborough and the crowds were treated to the sight of two Vulcans flanked by four Avro 707s in a fly past. Top to bottom: Avro 707A WZ736, 707B VX790, Avro 698 VX777 Vulcan prototype, Avro 707C WZ744, Avro 698 VX770 Vulcan prototype and Avro 707A WD280. In all, five small Avro 707 series deltas of varying designs were constructed, each powered by a 3,500lb s.t. Rolls-Royce Derwent engine and they provided valuable research to the Avro 698 programme before the bomber prototype flew in 1952. The first Avro 707 crashed on 30 September 1949. The prototype 698 (VX770) first flew on 30 August 1952. It was lost in a crash at Syreston during the 14 September 1958 Battle of Britain display. *Hawker Siddeley*

Vulcan prototype 698 (VX770) breaking up over RAF Syreston during the 14 September 1958 Battle of Britain display. All four crew were killed. *RAE*

Competition, which involved making three simulated atomic bomb attacks on a single night on San Jose in California and Boise and Batte in Montana, followed by an 800-mile cross-country route using astro navigation. In 1959 Valiant crews of No.214 Squadron established four unofficial long distance records, from Marham to Nairobi, Salisbury, Rhodesia, Johannesburg and from Heathrow to Cape Town. On 9 July 1959 a Valiant of No.214 Squadron made the first non-stop flight from the UK to Cape Town covering a distance of 6,060 miles after being twice refuelled in the air. In May 1960 a Valiant, again from No.214 Squadron, made a non-stop flight from the United Kingdom to Singapore after twice being refuelled in the air by Valiant tankers operating from Cyprus and from Pakistan.

In a portent of things to come, on May Day 1960 a Central Intelligence Agency (CIA) sponsored Lockheed U-2B high-altitude strategic reconnaissance aircraft piloted by Lieutenant Francis Gary Powers was shot down near Sverdlovsk by an SA-2 SAM on a flight over the Soviet union from Peshawar to Bødo in Norway. The U-2 was capable of operating over the Soviet Union at altitudes above 70,000 feet and it had seemed immune from interception. For almost four years, starting in July 1956, Twenty-nine previous secret missions had gone unchallenged by the Soviets but the new air-defence weapon meant a complete change in the role of the V-Bomber force. It now became clear that in the absence of a long-range stand-off weapon any air attack would have to fly at low level under the radar to their targets, but this was still four years away. In the meantime, V-bomber crews continued training as if little had changed. In early 1960 at No.232 OCU at RAF Gaydon, Chief Instructors fondly told new intakes that the Valiant was 'a gentleman's aircraft'.

Alan Gardener, who had been a Canberra navigator before flying in Valiants and who later served on Victor tankers at Marham, recalls: 'We formed No.57 Valiant Course, so we were certainly no pioneers, the first aircraft having been delivered to No.138 Squadron five years before. In those days the requirements in terms of flying hours, experience and "assessments" ensured that the V-Force really did represent the "cream" of Bomber Command, if not the Royal Air Force. Although I had flown 900 hours on Canberras, my only experience of radar bombing – my new "trade" – was gained at the Bomber Command Bombing School at Lindholme. We trundled around the sky in a fleet of Lincolns whose wartime H2S sets functioned (usually) thanks to what must have been superhuman efforts by the radar mechanics. Powered by four 10,000lb thrust Rolls-Royce Avons, the Valiants were able to carry a 10,000lb nuclear bomb on a 3,500-mile sortie, cruising at 0.76 mach at 35–40,000 feet. The Valiant was a nice big, comfortable aircraft to fly in. The crew compartment shared its general layout with the other V bombers. The first pilot (Captain) and co-pilot sat side by side in ejection seats, with fully duplicated controls and instruments. Behind them and facing aft, the Nav (Radar), Nav (Plotter) and Air Electronics Officer shared a table that spanned the full width of the cockpit, while facing them were the controls and instruments appropriate to their trade. Heart of the weapon delivery system and good enough to remain in service for over 25 years, was the Navigation and Bombing System (NBS), which allied a greatly updated H2S Mk.9 radar with NBC, an electro-mechanical computer of startling ability. With this equipment, the Nav, could, by placing the electronic markers on his radar screen over suitable responses, put a "fix" into his Ground Position Indicator, "home" the aircraft to a selected point or, if that point were to be a target, "fly" the aircraft on its bombing run. Bomb door opening and bomb release was achieved automatically.

But all this was several months in the future and we spent the spring in lectures on the aircraft systems, equipment and procedure, emergency drills, flight simulators and ground training aids of all descriptions until we were pronounced ready to be let loose on the real thing. Meanwhile we had formed

into a crew and one that in fact would stay together for the next two years, so it was lucky that we found ourselves quite compatible. At the end of May, a crew coach whisked us out to the dispersal, and up to the beautiful white bird where we clambered up the ladder and in through the big oval door in the side of the crew compartment. Only our co-pilot missed out on this familiarisation trip, his place being taken by an instructor. As we climbed in, I saw that a Spitfire was landing; it seemed a good omen. The Valiant was a nice big, comfortable aircraft to fly in and altogether we flew fourteen exercises in the next six weeks. There were cross-countries, day and night practice diversions to other airfields, emergency drills, instrument rating tests; and most trips included at least one example of something with which I was to become very familiar during my time on the V-Force, the Radar Bomb Score run.

Scattered around the country were seven or eight Radar Bombscore (RBS) Units, mostly located on airfields near to large towns. Within a radius of about 25 miles of each was a selection of targets, which it was the job of the Nav radar to attempt to "hit". These targets consisted of such things as bridges, factory buildings, and even crossroads, not necessarily visible on the radar screen. Fortunately, the designers of the Navigation and Bombing System equipment had foreseen this and provided an "offset" facility. So that by setting-in the distance in yards N/S and E/W of a known radar response from a target, and putting this under the markers, a decent score could be achieved. All this ensured that the Nav Radar was seldom at a loose end when not flying, as the choosing and measurement of offsets from all the published targets at each RBS unit provided many hours of harmless fun. As did the drawing up of overlays, drawn on acetate sheet to represent the expected response from the area of the target and laid over the radar screen to locate it. Photography of the screen was a useful backup should the RBS unit fail to track the aircraft on its run. This enabled an approximate result to be claimed for the crew and a library of photographs to be built up by the nav radars; those of the "old hands" being eagerly sought by the new boys.

Cross-country navigation was the province of the nav plotter, whose Ground Position Indicator (GPI) was fed with ground speed and drift from a Green Satin Doppler system. Updated fixes from the radar/radio fixes, or astro shooting the stars (or by day, the sun) was another little job for the nav

Vulcan B.2 XL320. *BAe*

First production Victor B1 XA917, which first flew on 1 February 1956. *BAe*

Victor tanker near Leuchars, Scotland in 1967. *Brian Allchin*

radar. A measure of the cavernous accommodation was that to reach the periscopic sextant, mounted in the roof, it was necessary to stand on the table. In true service tradition, whatever good equipment was fitted to an aircraft, procedures were evolved and practiced on the assumption that it would malfunction, be it the engines, flying controls, pressurisation system or any of the many bits in the navigation and bombing systems.

Before the end of July we passed out from Gaydon. We were posted to No.7 Squadron, recently transferred from Honington to Wittering. Like all the Valiant bases it was under the control of No.3 Group with headquarters at RAF Mildenhall. No.138 Squadron, the first to convert to the Valiant, had been in residence for five years. The third squadron, No.49, with only a slightly shorter tenure, had provided the aircraft and crews which dropped the first British hydrogen bomb at Christmas Island in 1957; quite a hard act to follow!'

The 45th and last Vulcan, B.Mk.1 XH532, was delivered on 30 April 1959 but this aircraft was much improved over the earlier machines. These and later improvements were refitted into the earlier aircraft with the main additions being a higher powered version of the Olympus engine, a large probe to allow in-flight refuelling and what was prob-

ably the most important addition, a large tail fairing containing electronic counter measures (ECM) equipment. This installation brought about the aircraft's re-designation to the B.1A. The prospect of stand-off weapons, advanced avionics and the enormous development potential of the Olympus engine brought about an extensive redesign of the Vulcan. The second prototype (VX777) was converted to Mk.2 standard, the most noticeable feature being a new, thinner wing of greater span and greater chord in the outboard sections, achieved by using new outer-wing panels, which added 12 feet to the 99 feet wing span. The B.2 was powered by 17,000lb thrust Olympus 200 series engines, later versions being fitted with the Olympus 300 engine of 22,000lb thrust. It had a service ceiling of about 65,000 feet and a greater load carrying capability than the B.1. The B.2 entered service with No.83 Squadron 1 July 1960 when XH558 was delivered to Waddington to 'B' Flight of No.230 OCU. The Waddington and Cottesmore Vulcan wings were powered with Olympus 300 series engines. The Scampton wing's Vulcan's were powered by Olympus 200s and armed with WEI77B free-fall nuclear weapons and 1,000lb iron bombs. Most B.2s were equipped to carry the Avro W100 'Blue Steel' stand off missile which originated in the 1950s in response to an Air

Avro Vulcan B.2 XL361 of 617 'Dam Busters' Squadron in 1962. *Brian Allchin*

Vulcan B.1 XA911 of No.230 OCU at RAF Finningley on 16 September 1961.
Tom Trower

Valiant B.K.1 XD866 at RAF Middleton St. George on 14 September 1963.
Tom Trower

Staff operational requirement. This rocket-propelled stand-off weapon was designed to be air launched a hundred miles from its target and Vulcan Mk.1 XA903 together with a number of modified Valiants was used as a development aircraft for the project. In 1966 Vulcans became operational in the low-level role, carrying up to twenty 1,000lb-gravity bombs or a nuclear payload up to a maximum range of about 4,600 miles.

Vulcan fuel-receiving training began late in 1960 in preparation for a non-stop flight to Australia the following June. On 20/21 June 1961 a Vulcan B.1A of No.617 'Dam Busters' Squadron at RAF Scampton made the first non-stop flight from the UK to Australia, covering 11,500 miles in just over 20 hours. The B.1A was flown from England to RAAF Richmond near Sydney, refuelled three times by No.214 Squadron's Valiants stationed along the route at Cyprus, Karachi in Pakistan and Tengah, Singapore. This, and other long-distance flights by Vulcan squadrons, demonstrated the global mobility and extended range now made possible. In 1962 No.617 Squadron became the first V-bomber squadron to become operational with in-flight refuelling. In 1963 Valiant tankers refuelled three Vulcan B.1As of the Waddington Wing over Libya, Aden and the Maldives for the flight to Perth, Australia and back.

In the meantime the Victor B.Mk.2 version was produced. The B.2 was a development of the Mk.1 and Mk.1A, powered by four 22,500lb thrust Rolls-Royce Conway RCo 17 Mk 201 engines. It differed exter-

nally from the B.1 in having wings of greater span, enlarged air intakes, a dorsal fillet forward of the fin and two retractable scoops near the tail to supply ram air to two turbo-alternators for emergency power supplies. Like the Vulcan B.2, the Victor B.2BS could carry the Blue Steel stand-off bomb. The B.Mk.2 prototype flew in February 1959 and after problems with engine compressor surging were eliminated, John Allam and his colleagues 'settled down into its trials programme with enthusiasm'.

By mid-August the aircraft had completed 98 hours' flying when it was handed over to the Aircraft and Armaments Experimental Establishment (A&AEE) for their initial appraisal. Prior to its hand-over the engine problems had been resolved, the aircraft had been flown to nearly 58,000feet, manoeuvre margins had been extensively investigated up to 55,000 feet and IMN 0.95, auto-stabilizer functioning had been checked and Mach trimmer runways investigated. Climb performance was magic by comparison with the BMk1. It really was in good shape and ready for its Boscombe Down appraisal. Sadly, the aircraft did not return from its first flight in the hands of the A&AEE crew and was subsequently discovered to have crashed into the sea off the south-west coast of Wales. Although the second B.Mk.2 had already flown, the loss of the first was a major setback. However, flight testing with the second and subsequent B.Mk.2s continued at Radlett and despite further intensive investigation into every aspect of the aircraft,

Avro Vulcan B.2 XM651 of No.101 Squadron from RAF Waddington at RAF Cottesmore on 14 September 1968. *Tom Trower*

including high-altitude and high IMN handling, nothing came to light to indicate a possible cause for the loss of the first B.Mk.2. Other aspects of the BMk2 trials continued and included performance, autopilot and later autoland, Blue Steel carriage and release (at Woomera in Australia), and trials associated with the new electrical system including use of the ram air turbines to operate the flying controls. The first Victor B.Mk.2s delivered to the RAF formed the B.Mk.2 Trials Unit at Cottesmore in September 1961. In February 1962 No.139 (Jamaica) Squadron started to equip with B.Mk.2s as also did No.100 Squadron in May 1962. B.Mk.2 developments continued after the aircraft's entry to service and these included the fitting of the definitive R.Co17 (Rolls) engines, upgraded ECM equipment, rapid take-off capability, and the fitting of 'Kutchemann carrots' or 'Whitcomb bodies' to the upper wing surfaces.

At Wittering during the winter 1960-61 Valiant navigator **Alan Gardener** on No.7 Squadron noted that: 'Diversions became a not unfamiliar feature of life, our Captain's White Instrument Rating preventing us "getting back in" when the weather got below his limits. But at least the procedures and letdowns at unfamiliar airfields were valuable practice.

No.3 Group decreed a series of Basic Training Requirements to be achieved in each six-month period for each member of the crew. These included a specified number of Instrument Landing System approaches, asymmetric landings or various practice emergencies for the pilots, navigation stages of different types and bombing scores for the navigators. In parallel with this, the achievement of a consistent series of scores in navigation, and bombing, would lead to a grading of the crew as "Combat" and subsequently "Select", while for the top crew the designation, "Command" was given. We achieved Combat by the end of March 1961, which made us eligible for two new experiences. The first was to go on the list for the next "Lone Ranger" overseas flight and the other, slightly less welcome was to be put on the Quick Reaction Alert (QRA) roster; we were at last a real part of the deterrent force!

The three squadrons each held one aircraft and crew on alert at all times, the aircraft armed with a Blue Danube weapon and "cocked" or "Combat Ready" in correct parlance. In normal use, from climbing in to taking off would take 45 minutes. But if a crew checked each item of equipment in similar fashion to a pre-take-off check and left all switches in specified positions, it was possible to reduce this to about 5 minutes on a good day.

The alert crew was required to stay together throughout its tour of duty. Flying clothing was worn, meals were taken in the aircrew buffet and the day spent in activities that could be immediately abandoned if a call to "cockpit readiness" was "tannoyed". It was an appropriate time to spend the passing hour or two studying the allocated war targets and planning routes. The crew's "go-bag" was drawn from the Wing Weapon's Leader and the crew locked in the 'Vault' with suitable intelligence and briefing material. Outside the confines of the "Vault", one member of the QRA crew would have the "go-bag" with its contents, locked to his wrist at all times. At night, although originally allocated dedicated adjacent rooms in the Mess, it was obviously decided that too much comfort was a bad thing for we were later provided with a "QRA caravan" for sleeping purposes. This was a little like a miniature railway carriage (suburban!) with five individual compartments each complete with bunk, wash-basin and heater, but not much else. Three of these devices were parked adjacent to the Ops Block. QRA could be an irritation; but at a weekend, when the rest of the station was away and gone, and the only people to be seen were those other poor souls on duty of one sort or another, it was a real pain. One of the techniques planned for our survival as a reprisal force in the event of a Soviet Nuclear strike was for all available V-bombers to disperse, in groups of four, to two dozen airfields around the British Isles. These were equipped with suitable accommodation, hardstandings, communications, Motor Transport and in most eases end-of-runway Operational Readiness Platforms (ORPs), to function as emergency strike bases should the need arise.

Vulcan B.2 XH536 at Farnborough on 11 September 1959.
Tom Trower

The need arose in May, when Exercise Mayflight was called, and we found ourselves camped in a far corner of Lyneham. After three days on various states of readiness and general turmoil, we finally scrambled under orders from the Bomber Controller, through our secure tele-scramble link, and sortied back to Wittering. Ten days later we departed on our first "Lone Ranger". A whole day in sunny Bahrain, with night stops in El Adem, a busy RAF staging post on the route to the East. Accompanying us in the sixth seat was the crew chief, whose job it was to supervise the turn-round servicing and refuelling wherever we landed. Should there be no RAF ground tradesmen, he would supervise us, as we all carried certificates of "Competence to undertake first line servicing" in specific trades.

Amazingly, only two weeks later we were off again, this time to the American mid-west Offutt Air Force Base, the HQ of Strategic Air Command in Nebraska. This huge base was a source of great fascination, with its acres of aircraft parking, B.47 bombers and EC-135 command posts of the "Looking Glass alert" force being among many types to be seen. The staging on outward and return flights was through Goose Bay in Labrador, another huge base, Even on the far side of the Atlantic we couldn't escape the RBS Units and three targets were "attacked" in North America in each direction. On our return, we began planning for yet another Ranger flight a week later, this time a very rare opportunity to visit Salisbury in Southern Rhodesia, now Zimbabwe. We routed through Idris and Nairobi, a round trip of more than 10,000 miles and a most absorbing contrast to our American sortie. The remainder of 1961 was passed in routine flying, cross-countries with RBS attacks and occasional live bombing on the Jurby or Wash ranges, using 100lb practice bombs aimed either by radar, or visually when the Nav Plotter would clamber down below the floor and lie in a small cupola equipped with a T.4 bombsight. Also, Group and Command exercises were also flown, as were Instrument Ratings, air tests and circuit training, with bouts of QRA from time to time to add variety.

One further Western Ranger took place in November. By this time the continental winter was well in evidence, and the sub-zero temperatures at Goose Bay meant that the aircraft would be parked overnight in the huge, heated RAF hangar. In the morning, all the checks up to engine start were carried

Avro Vulcan armed with 'Blue Steel' taking off.
Avro

Vickers Valiant B.(K).1 XD816. *MoD*

Blue Steel unit, No.100 Squadron, formed in May, taking over No.138's accommodation. Normal activities continued during the summer, the unit taking part in several exercises, while weapon trials on a new weapon, Red Beard, were carried out. A Lone Ranger flight to Aden came our way in July, routeing through Cyprus outbound and Malta on the return flight, all bases fully furnished at that time with RAF servicing flights for transit aircraft. September was to be the final month of No.7 as a Valiant squadron, a full flying programme of cross-countries and RBS runs being maintained, our last sortie being to ferry a Valiant to Gaydon for use on the OCU. On 1 October the squadron disbanded, when No.27 Squadron with its new Vulcan B.2s at Scampton took over the Alert commitment and war targets.'

During 1961–62 proposals were made for two- and three-point Valiant tankers, but these were turned down in favour of a three-point installation in the Victor. A number of Mk.1 and IA Victor bombers became available as squadrons re-equipped with the Victor B.Mk.2. Three squadrons of Victor tankers were required to supplement the Valiant force and eventually replace it in the late 1960s and Treasury approval for the conversion was obtained in 1963. Following the discovery of fatigue defects, however, the Valiant force had to be withdrawn from service in January 1965 and the Victor conversion programme was then accelerated to make good this sudden loss of tankers. The prototype conversion (XA918) flew in September 1964 and was used to test the compatibility of all current receiver types.

In January 1964 the Victors, Valiants and Vulcans were finally switched to the low-level role with conventional gravity bombs or a nuclear payload and they began to appear in a new camouflage paint scheme appropriate to this new mission profile. However, problems soon arose with the Valiant. This aircraft had been designed for high-level, long-range operations but in 1960-61 Nos 49, 148 and 207 Squadrons had been assigned to NATO in a low-level tactical

out before the hangar doors were opened and we were towed outside to start up and continue on our way. At Offutt, a little before our departure time, we heard the sound of crash vehicle sirens, and going outside saw a rising cloud of black, oily smoke. What a realistic practice, I thought at first, but it wasn't. A Wyton Valiant, taking off an hour before us, had failed to get airborne, slid down a steep embankment and across a main road at the end of the runway, the crew compartment breaking away from the rest of the aircraft. With great good fortune all escaped without serious injury. Failure to properly de-ice the aircraft was blamed for the accident.

During 1961 No.49 Squadron departed for Marham and we became aware of new building and equipment at Wittering in preparation for future developments. This translated early the next year into the formation of No.139 Squadron and the arrival of their exciting new Victor 2s, equipped to carry the Blue Steel stand-off missile. In March, having achieved a "Select" category we were allotted another Western Ranger to Offutt. On return, an unusual trial, or perhaps a way of disposing of unwanted ordnance, was carried out on the range at Luce Bay by the Mull of Galloway, when we were tasked to drop the Valiant's full conventional load of twenty-one 1,000lb bombs, an interesting experience. The following month saw the disbanding of No.138 Squadron, leaving No.7 Squadron to maintain QRA duties alone, with the second and final Victor 2

Line up of Victor bombers. *MoD*

bomber role. Stresses on the airframe became too much and serious metal fatigue in the rear wing spars led to the Valiant fleet being grounded on 11 December 1964. Consequently, the following January all Valiants were withdrawn from service and subsequently scrapped. Altogether, 104 aircraft had been built and had equipped ten squadrons in Bomber Command. At Marham on 1 March 1965 No.207 Squadron disbanded and the two others followed on 1 May. The last production Vulcan was XM657, which was delivered to No.35 Squadron on 14 January 1965. Four years later, on 30 June 1969, the RAF relinquished the strategic QRA deterrent role to the Royal Navy's Polaris nuclear submarine fleet and Vulcans were converted to the free-fall bomber role. Vulcans were deployed to Cyprus to help strengthen Central Treaty Organization (CENTO), with Nos 9 and 35 Squadrons sending four aircraft each from Cottesmore during January. The Valiant had been designed for high-level, long-range operations but in 1960-61 Nos49, 148 and 207 Squadrons had been assigned to NATO in a low level tactical bomber role. Stresses on the airframe became too much and serious metal fatigue in the rear wing spars led to the Valiant fleet being grounded on 11 December 1964–February 1969 to Akrotiri,

the squadrons vacating Cottesmore completely by March 1969.

Meanwhile, the decision had been taken for another former V-bomber to take over the role of aerial tankers. In January 1965 the Handley Page Company had begun working around the clock to convert six Victor B.1As to B(K)2A two-point tankers. That same month No.92 Squadron's Lightnings were forced to 'puddle-jump' their way from the UK to their detachment in Cyprus via West Germany, southern France, Sardinia, Malta and Libya. Meanwhile, the conversion of Victors to in-flight refuellers was speeded up. The first to fly was XH620 on 28 April. On 24 May No.55 Squadron, which had become non-operational as a Medium Bomber Force Squadron at Honington on 1 March, moved to Marham to operate in the in-flight refuelling role. Ironically, Second Lieutenant (later Sir Alan) Cobham, the great air-refuelling pioneer, had been one of the flying instructors at Narborough during the last two months of World War One.

John Allam recalls: 'With the entry of the Victor B.Mk.2 into service it was planned that the B.Mk.1s would be withdrawn from the V-Force and modified to become flight refuelling tankers to replace the Valiant tanker fleet. The abrupt end of the Valiant brought this programme forward and put

In 1959 No.214 Squadron operated the B(PR)I multi-purpose version of the Valiant for flight-refuelling trials. Valiant B(PR) K.1 WZ376 and B(PR)1 WZ390 demonstrate air-to-air refuelling. *Vickers*

Valiant B(K)I XD815 of No.90 Squadron at Khormaksar, Aden in 1964. *Ray Deacon*

Victor B.1A XH650 of No.55 Squadron at Khormaksar, Aden in 1963. *Ray Deacon*

considerable pressure on Handley Page to provide Victor tankers very quickly. This was achieved by modifying six B.Mk.1s to two point K.Mk.2 instead of the definitive arrangement as three-point tankers. As a result the gap between the Valiant tanker demise and the entry of Victor tankers to service was minimal. Tanker trials commenced at Radlett in September 1964 and went through quickly and with very little trouble to obtain service release, so that the first Victor tankers were able to enter service in May 1965. Not long after the B.Mk.2 became operational in service it was apparent that altitude would no longer provide the defence that had been enjoyed thus far because of the development of the surface-to-air missile. The technique therefore had to be changed to the low-level bombing role, enabling the aircraft to fly below long-range radar cover to the target. As the new aircraft specifically designed for this role were not yet ready, the V-Force aircraft were suddenly required to operate at low level. Hence terrain-following radar was fitted and the aircraft started to operate in their new environment. Whilst the Victor, from the handling and performance point of view, was perfectly capable of fulfilling this new task, its airframe had never been designed to

spend long periods of time at comparatively high Indicated Airspeed (IAS) in predominantly turbulent air. It was a lightweight airframe intended for flight at high altitude in calm air and hence, at low altitude, its fatigue life began to be devoured at a rather high rate. However, the B.Mk.2s continued in their bombing role until they were finally phased out at the end of 1968. The aircraft went into storage at Radlett awaiting their conversion to the tanker role. No.543 Squadron's eight SR.Mk.2s were retained as radar reconnaissance aircraft and continued in this role until 1973.

The Victor SR.2 was the strategic reconnaissance version and first entered service with No.543 Squadron at Wyton in late 1965. SR.2s differed from previous versions in having a bomb bay converted to accommodate camera equipment and additional fuel tanks for increased range. During November 1973 some of the Vulcan's began to take on yet another role with the establishment of No.27 Squadron, as a maritime radar reconnaissance unit having been removed from the strike role in March 1972. The Vulcans converted for this new role became known as SR.Mk.2s and No.27 Squadron took the place of No.543 Squadron's SR Mk.2 Victors when it disbanded in May 1974.

Over the Alps Lightning F.3 of No.29 Squadron moves in to refuel from Victor K.1A XH649 of No.57 Squadron during an exchange exercise to Grosseto, Italy in 1970. *Dick Bell*

The B.Mk.2s remained at Radlett whilst the Ministry procrastinated about a start date for their conversion to tankers. Largely as a result of this indecision Handley Page went out of business at the end of February 1970 and within a week the contract for the conversion of these aircraft was awarded to Hawker Siddeley. Arrangements were then made to fly all the Victors standing at Radlett to Woodford where the conversions would be made. These aircraft had been at Radlett for over a year. Work immediately started to service them for one flight to Woodford. The first departed from Radlett on 25 March 1970 and the process continued in a steady stream with the final one leaving on 10 July. There was then a deathly hush whilst Woodford set itself up to undertake the conversion of the B.Mk.2s to K.Mk.2s – a job which could have been completed at Radlett in a very short time. It was not until May 1974 that the first Victor K.Mk.2 was delivered to the Air Force.'

During 1965–67 all Victors were converted to K IA air-to-air refuellers in the air-tanker role, and were equipped with up to three refuelling points for fuel transfer. No.55 Squadron's first two Victor tankers (XH602 and XH648) arrived at Marham on 26 May 1965 and in July a Tanker Training Flight was formed at the station. The two-point tanker was an interim measure only and was to be withdrawn and replaced with the more versatile three-point tankers when available. Two-point tankers did not permit Victor to Victor refuelling operations but in August they provided the air-to-air refuelling element for No.74 'Tiger' Squadron's Lightning F.3's deployment to Akrotiri, Cyprus. Meanwhile, six Lightning pilots from No.19 Squadron at Leconfield converted to high-level in-flight refuelling and then carried out low-level in-flight refuelling trials from five of the K.1A tankers in preparation for No.74 Squadron's return to the UK. Low-level tanking was possible but extremely difficulty

and was best carried out over the sea where conditions are relatively smooth. The Victor's wing was prone to flexing in turbulence and the whip effect of the basket end of the drogue could be quite 'frightening' if a Lightning was not on the end to 'tone it down'.

No.55 Squadron's full complement of six Victor tankers was finally reached in October. That month four Lightning F.3s of No.74 Squadron were able to use all six Victor tankers at Akrotiri in Operation Donovan when they were refuelled all the way to Tehran and back. On 1 December 1965 No.57 Squadron moved to Marham from Honington but the first of their six Victor K.1/1A three-point tankers did not arrive until 14 February 1966. That same month No.55 Squadron began receiving the first of fourteen Victor K.1A tankers in place of their two-point tankers. On 1 July No.214 Squadron re-formed at Marham as a Victor tanker squadron and by the end of the year was equipped with seven K.1/1A three-point tankers. In February 1967, Exercise Forthright 59/60 saw Lightning F.3s flying non-stop to Akrotiri and F.6s returning to the UK, refuelled throughout by the Victor tankers. In April No.56 Squadron's Lightnings flew to Cyprus carrying out six in-flight refuellings for the fighter versions and ten for the two-seat T-birds. In-flight refuelling from Victors also became a feature used often on QRAs in the UK. At night floodlights aboard the Victors had to be turned off to prevent blinding the Lightning pilots as they tanked over the North Sea. In June 1967 seventeen Victor tankers refuelled No.74 Squadron's thirteen Lightnings which flew to Tengah, Singapore in Operation Hydraulic, the longest and largest in-flight refuelling operation hitherto flown. All the Lightnings reached Tengah safely, staging through Akrotiri, Masirah and Gan. The Lightnings remained at Tengah for four years and during this time three 2,000-mile deployments were made to Australia non-stop using Victor tankers, the

A Victor K.1A refuels a Lightning from the centreline tank during a tanking exercise in the summer of 1972.
Dick Bell

major one being Exercise Town House 16–26 June 1969. Meanwhile, on 29 November 1967 a No.11 Squadron Lightning established a record of eight and a half-hours' flying, refuelled five times, flying 5,000 miles.

By now a further twenty-four Victor K.1/1A three-point tanker conversions were in RAF service. Operating from hot and short airfields, the Victor K.1 tankers were limited in the amount of fuel they could carry and it was sometimes necessary to refuel the primary tanker from another tanker before proceeding to the rendezvous. A new tanker having greater capacity and more power was required. Plans were made for the conversion of twenty-nine (later reduced to twenty-four in 1975) Victor B.Mk.2 and SR.2 versions, the type having been phased out of service in the bomber role when the task of maintaining the nuclear deterrent passed to the Royal Navy's Polaris submarine fleet in 1969. Preliminary design for the conversion was initiated by Handley Page, but after the collapse of this company, Hawker Siddeley Aviation revised the design and completed the conversions at Manchester, the first flying on 1 March 1972. No.55 Squadron re-equipped early in 1976 followed by No.57 Squadron in June but No.214 Squadron, the last to operate the Mk.1 tankers, was disbanded in January 1977 as a result of National Defence cuts. Designed for a 14-year life the Victors were due for replacement by the early 1990s. The RAF planned to bring nine converted VC 10s into service to supplement the fleet.

By the end of 1968 both Nos 100 and 139 Squadrons had been disbanded and sixteen of their Victor aircraft were subsequently modified to K.2 standard, replacing the K.1 tankers at Marham. The withdrawal of British forces from the Far East involved a rapid reinforcement commitment. In May 1968 four F.6 Lightnings of No.5 Squadron at Leconfield flew non-stop from Binbrook to Bahrain in eight hours, refuelled along the 4,000-mile route by the Victor tankers. Deliveries of the Victor K.Mk.2 began in May 1974 to No.232 OCU at Marham. This unit now took over from the squadrons, full role conversion training, in addition to ground training in a simulator and the Air-to-Air Refuelling School. Night and day sorties were flown when weather permitted. The rapid reinforcement commitment continued. On 6 January 1969 Exercise Piscator, the biggest in-flight refuelling exercise so far mounted by the RAF, took place when ten Lightning F.6s of No.11 Squadron, refuelled by the Victor tankers, deployed to Tengah staging through Muharraq and Gan and back, a distance of 18,500 miles. Victors of Nos 55, 57 and 214 Squadrons refuelled each Lightning thirteen times. Throughout the two-way journey 228 individual refuelling contacts were made during which 166,000 imperial gallons (754,630 litres) of fuel were transferred. In-flight refuelling was used for a less belligerent role in the 1969 Daily Mail transatlantic air race, when Harriers of Air Support Command and Phantoms of 892 Squadron, Royal Navy, were refuelled numerous times by Victor tankers based on both sides of the Atlantic. During Christmas 1969 Exercise Ultimacy, involving ten F.6s, of No.5 Squadron flying to Tengah for joint air-defence exercises in Singapore, only one stop, at Masirah, was made en route, the first Lightnings leaving Binbrook before dawn on a foggy 8 December morning. At 0345 hours they made their first rendezvous with their Victor tankers over East Anglia in the dark. Then it was on, across France and the Mediterranean and a rendezvous with more Victors from Cyprus. They continued eastwards, finally crossing Muscat and Oman and on to Masirah, the overnight stop. Next day a pre-dawn take-off saw the Lightnings away on the last leg of their trip, across 4,000 miles of ocean, south-east to Gan, where Victors rendezvoused on schedule to refuel them, and then due east to Sumatra. Some of No.74 Squadron's Lightnings were in need of major overhaul and pairs of Lightnings left for the UK refuelled by the Victors on the homeward trip after stopovers at Masirah and Akrotiri en route.

In May 1970 three Phantom FGR.Mk.2s of No.54 Squadron participating in Exercise Bersatu Padu flew to Singapore non-stop in a little over 14 hours with the aid of nine in-flight refuellings. The run-down of the aircraft-carrier fleet meant that land-based strike aircraft had to operate further out to sea. In exercises in the mid-1970s, tanker support enabled RAF Phantoms to mount a defensive patrol over a naval task force 900 miles from their base in Scotland and Buccaneers from Honington to carry out attacks against simulated enemy shipping operating more than 1,100 miles from the coast. In February 1971 No.23 Squadron tanked from Victors to Cyprus. On 25 August No.74 Squadron disbanded at Tengah and No.56 Squadron acquired all the 'Tigers' remaining F.6s, which, starting on 2 September, were flown over the 6,000-mile route, a 13-hour trip, to Akrotiri, staging through Gan and Muharraq and completing seven air-to-air refuellings with Victor tankers.

Opposite page:
Victor K.1A about to refuel a Lightning of No.19 Squadron. *Dick Bell*

Victor K.1A XH651 refuelling a Lightning of No.19 Squadron. *Dick Bell*

A Victor K.1A of No.57 Squadron and Lightning T.5 XS458 during air-to-air refuelling. *Dick Bell*

Victor K.2 XM717 'Lucky Lou' of No.55 Squadron. *Author*

Above and below:
Victor K.2 XL231
'Lusty Lindy' of No.55
Squadron. *Author*

Demonstrating the rapid deployment made possible by in-flight refuelling, in 1974 six Lightning F.6s were sent to Cyprus when a Greek-led coup by the Cyprus National Guard overthrew President Makarios. In July twelve armed Phantoms were dispatched from RAF Coningsby at short notice during the Turkish invasion of the northern part of the Mediterranean island. Flying through the night and refuelled by Victor tankers from Marham (which then landed at Malta), the fighters were available to the United Nations commander by first light the following morning. On 28 January 1977 No.214 Squadron at Marham disbanded for the last time leaving just twenty-four K.2 tankers of Nos 55 and 57 Squadrons as the only aircraft capable of in-flight refuelling the RAF's Lightning, Phantom, Buccaneer and Jaguar strike aircraft. (It was not until 1978 that it was announced that a number of VC 10s were to be brought back into RAF service for conversion to tanker aircraft to supplement the Victor K.2 force).

The rundown of the eight Vulcan B.Mk.2 squadrons, based within sight of Lincoln Cathedral since 1975 in the forefront of RAF Bomber Command's strike force (Strike Com-

mand after 1968) began in 1981 with the Panavia Tornado destined to replace the Vulcans in total. First to go was No.230 OCU in August 1981, followed, on 22 December by No.617 (but soon re-formed on the Tornado GR.1) and No.35 Squadron in February 1982. Waddington was expected to continue to fly Vulcans for a couple of years, until the Tornado was firmly established in service, but in November 1981 it was revealed that No.9 Squadron would be disbanded in April 1982, followed by Nos 44, 50 and 101 by 30 June. At Scampton No.27 Squadron had used the Vulcan SR.2 strategic reconnaissance version in the maritime radar reconnaissance (MRR) role since 1973 and on 31 March 1982, disbandment day, the Squadron flew its last Vulcan sortie and handed over to the Nimrod fleet. On 25 April the first eight Tornado GR.1s of No.617 Squadron arrived at Marham to share the station with the in-flight refuelling crews of the Victor tankers. The Victor fraternity was a closely-knit family who took great pride in their exclusive role and unashamedly referred to themselves as the 'Tanker Trash'. Once the Victors (and the Vulcans) had been a part of the V-bomber elite and soon the old guard was referring to No.617 Squadron as 'The Dim Bastards' and, according to the Victor crews, MRCA stood for 'Much Refurbished Canberra Aircraft'. While the 'Dam Busters' worked up on the Tornado at Marham it was the 'Tanker Trash' and the few equally elderly remaining Vulcans that were suddenly and unexpectedly thrust into the ascendancy again.

On 1 April South Georgia and the Falklands were invaded on the orders of the Argentine Junta in Buenos Aires and the veteran Victors' and Vulcans' presence in the South Atlantic was urgently required as part of the British operation to regain the islands, code-named Operation Corporate. On 18 April the advance party and five Victor K.2 tankers left Marham for Wideawake airfield on the British-owned island of Ascension, 3,375 miles from the Falklands. Next day four more Victors flew the 4,100 miles from Marham to Ascension and by the end of the month fourteen Victors were stationed at Wideawake.

The Nimrod force, unlike the Victors, did not have the refuelled range to operate as far south as the Falklands so on 20 April Squadron Leader John Elliott and his crew in XL192 flew the first operational Victor sortie during Operation Corporate. The primary purpose of the MRR (and two similar MRR sorties flown by Victors on 22/23 and 24/25 April) was to provide data on surface shipping and ice conditions etc to HMS *Antrim*, the ship leading the small naval Task Group responsible for recapturing South Georgia (Operation Paraquat). Having reached its target area Elliott descended from his transit height to around 18,000feet and for 90 minutes Beedie and Cowling carried out a radar sweep of 150,000 square miles of ocean. Nothing untoward was found and the information was made available to *Antrim*. The 7,000-mile sortie took 14 hours 45 minutes. South Georgia was recaptured on 26 April. On 29 April Vulcan B.2s XM598 and XM607 were tanked from Waddington to Ascension. The Victors air refuelled the Vulcans twice en route and the tankers required numerous fuel transfer themselves to complete the 9-hour flight.

Squadron Leader A.1.B. 'Al' Beedie of No.55 Squadron, who, with Squadron Leader Tony Cowling, formed the radar team aboard Squadron Leader John Elliott's Victor, recalls: 'Beginning on 20 April, Victors, supported by five more operating in the air refuelling mode, flew three maritime reconnaissance operations, each more than 14 hours duration, to waters in the region of South Georgia. On 30 April/1 May two Vulcan B.Mk.2s of No.101 Squadron, supported by ten Victors, set out on a demanding 7,860-mile round trip to bomb Port Stanley in an incident-filled, complex operation [Black Buck I]. At the start the primary Vulcan [XM598] aborted after its cabin could not be pressurized and the Victor fleet, too, suffered malfunctions. Victors refuelled Victors until only two remained with the reserve Vulcan [XM607], flown by Flight Lieutenant Martin Withers and Flight Lieutenant Dick

Russell. [Russell, who had just turned 50, was an experienced Victor AAR instructor at Marham]. Just before the fifth refuelling of the Vulcan one of the Victors [Squadron Leader Bob Tuxford flying XL189] attempted to refuel the other Victor in very turbulent conditions and the receiving Victor [XH669 flown by Flight Lieutenant Steve Biglands] had its probe broken during the transfer. The two Victors exchanged roles and the provider took back the fuel. Although dangerously low on fuel itself the Victor then transferred enough fuel to the Vulcan to allow it to make its attack, before heading back towards Ascension and calling for another tanker to meet it. Withers successfully dropped his twenty-one 1,000lb bombs on Port Stanley airfield. The Vulcan was then refuelled a sixth time and returned to Ascension after being airborne for 15 hours 45 minutes. At the time it was the longest-range operational bombing operation ever flown.

The last refuelling bracket took place 3,000 miles south of Ascension and Tuxford had to transfer more fuel than planned, which left XL189 with insufficient fuel to return to Wideawake. The need for radio silence meant that the Victor could not call Ascension and arrange for another tanker to meet him on the return. Tuxford and his crew could only pray that another tanker from Wideawake would intercept them before they ran out of fuel and had to ditch in the freezing South Atlantic when death would be almost instantaneous. Black Buck I was the first time that a Vulcan had dropped bombs in anger in its 25-year history.'

Squadron Leader Bob Tuxford of 55 Squadron recalls: 'We did a number of work-up sorties before leaving for Ascension. We had an inkling of what was ahead, but I think it's fair to say no one expected to be launching eighteen-ship sorties and assisting a Vulcan to drop the hardware. It was a massive operation. We were taken aback

Victor K.2 of No.55 Squadron with brake chute trailing, taxiing in at Marham in October 1993. *Author*

No.55 Squadron Victor line-up at Marham in October 1993. *Author*

with the logistical aspects. Air-to-air refuelling was our business, but normally escorting fighters around the world in much reduced numbers. The tanking support for this mission took eighteen individual Victor sorties – a very complicated operation, but one we thought absolutely achievable. We had a fuel planning cell at No.1 Group that looked after the tanker force. Mostly ex-navigators, their business was to produce fuel plans. Black Buck was just a variation on a theme. We saw the plan for the very first time on the evening of 30 April, but it was our experience using formation procedures that allowed us to run with it quickly. Martin Withers has acknowledged that the plan was mind-boggling to him. He had Dick Russell with him in the cockpit to bring some air-to-air refuelling expertise into the Vulcan crews. [The Vulcan crews had just 12 days in which to hone their refuelling skills.] I flew over 14 hours that night. I went through every emotion from excitement in the launch phase to worry in the early stages.

Much of the "excitement" for me came in the sixth and seventh hours when I was refuelling one of the other tankers, Victor K.2 XH669 piloted by Steve Biglands, which broke its probe in my basket. That necessitated changing places and receiving back the fuel we'd just passed in pretty shocking weather. The degree of difficulty at that stage was as hard as anything I'd had to deal with before. Despite a shortage of fuel in the whole formation, I was able to give the final offload to the Vulcan, which then went onto the Falklands, albeit with less fuel reserves than hoped. As I "turned the corner", it was seven hours back to Ascension and we had five hours worth of fuel. I didn't have any options to divert; there was no option but the South Atlantic. We were very focused at that point, not sure if the Vulcan could finish the job. But once we'd intercepted the code word that indicated the job was done we went from a subdued state to an elated one and then concentrated efforts on recovery, which would require another tanker. The original plan hadn't allowed for such a contingency so I was relying on their knowledge at Ascension on how the plan had gone and the fact that the formation was short of fuel, so they would launch additional recovery aircraft, which indeed they did.

The Vulcan crews were not very familiar with air-to-air refuelling operations – the managers at the Vulcan bases did not have the best available information to them on the fuel consumption rates the aircraft would endure, especially fully laden, and undertaking multiple formation changes. So, the figures were not as accurate, in retrospect, as we would have liked. A number of things also went significantly wrong, Steve Biglands' unfortunate breaking of the probe being the biggest. We were flying in towering cumulous clouds at night. Your visual references for formating on the aircraft in front are reduced. Therefore with the turbulence, the distracting lightning and St. Elmo's Fire all around the cockpit windows the whole process of achieving a stable contact and maintaining it for long enough to get the fuel on becomes much more difficult. On approaching the Falklands, the whole formation had burnt a lot more fuel than had been planned. Unfortunately, that ended up with me, the final tanker, 20,000lb short.

There is a weak link on the front end of the probe that is designed to shear if too much lateral force is put upon it. One of two things can happen; the probe tip can lodge inside the coupling of the basket, which would render it useless, or the tip breaks off and falls into the sea, if it doesn't enter the engine intakes. At this point, I wasn't aware whether I would be able to use the centre HDU [Hose Drum Unit] again. But everything premised on my ability to change places with Steve Biglands and take back the fuel we had just given him. If I couldn't do that, then it didn't matter that my hose may be damaged. So, I took back the fuel, so I had enough to go on with the mission. Then we had to decide if the hose would be satisfactory. We positioned Withers' Vulcan behind and visibly inspected it but that failed to provide sufficient confirmation, so we made a small additional transfer of about 5,000lb. That demonstrated how flexible this whole refuelling plan was. We were effectively making it up as we went along. Once I knew he could take on fuel, we could then continue with the mission.

If we hadn't been able to transfer to Withers, the mission would have been aborted. I had grounds for aborting the mission somewhat earlier, as I was prejudicing my own recovery because I wasn't sure we could meet another tanker on the way home. I hasten to add that I discussed the options available with my crew and each of the other four members came back and said. "Let's press on and get the job done". That was in the full knowledge that we would be two hours short of fuel in getting back to Ascension, about 600 miles south.

We gave the Vulcan sufficient time to effect the mission and my AO, [Airborne Electronic Officer] Flight Lieutenant Mick Beer, intercepted the code word "Polo" designed to communicate the mission had been a success. I was then able to make arrangements for my own recovery. Two Victors were scrambled from Ascension. They were from the recovery wave of six tankers for the mission to bring the Vulcan back post-strike. The six aircraft were required to provide two Victors at the Rendezvous (RV) in case of problems. Fortunately, Group Captain Gerry Price, the Station Commander from RAF Marham who was running the Air Bridge at Ascension made the decision to divert two of the Victors to assist us. Another of our tanker crews, Flight Lieutenant Alan Skelton, had developed a fuel leak and also required assistance. They met us three hours south of Ascension when we had about an hour's fuel left.

I'm very proud, both for my crew and the whole of the tanker force. Any one of those eighteen tankers failing would mean the mission failing. Pretty much the whole of No.1 Group's tanker force was on the island and it was thanks to some pretty skilful flying from every pilot that enabled the whole plan to be accomplished. It was very much more than a simple bombing mission. The Argentines then knew we could attack them on the mainland.'

Bob Tuxford was awarded the Air Force Cross for his part in the mission and his crew: 52-year old Squadron Leader Ernie Wallis MBE, Flight Lieutenant Mick Beer, Flight Lieutenant John Keable and Flight Lieutenant Glyn Rees, each received the Queen's Commendation for Valuable Service in the Air.

Black Buck 2 went ahead on 3 May and was flown by the same Vulcan with a different crew who took off from Wideawake with a small group of Victor tankers. A larger formation of Victors left later and flew at a higher speed than on Black Buck I to catch up with the Vulcan well along the route to Port Stanley. This operational change worked well enough but the runway was not hit though several Argentinean aircraft and buildings were very badly damaged. Black Buck 3 scheduled for 16 May was cancelled because of bad weather. Black Buck 4 flown on 28 May was the first anti-radar mission but five hours out from Ascension the HDU, the mechanism that winds and unwinds the refuelling hose, failed in the key Victor tanker and the mission was aborted. Two nights later, on 30/31 May on a mission to the Port Stanley area, Black Buck 5 supported by eighteen Victor sorties was successful and the Vulcan launched four AGM-45A Shrike missiles, which damaged the antenna of the Argentinean Westinghouse AN/TPS-43 surveillance radar installation. Black Buck 6 went ahead on 2/3 June but the Argentinians repeatedly switched off their radars. Vulcan XM607 was forced to divert to Rio de Janeiro in Brazil after an in-flight refuelling accident involving the loss of its probe on the return leg of the anti-radar strike. Black Buck 7 flown to bomb Port Stanley airfield on 11/12 June was the last bombing operation involving a Vulcan. Fourteen Victor tankers carried out eighteen refuelling sorties. The mission was a success with the Vulcan dropping a full load of 1,000lb high explosive and anti-personnel bombs on the airfield without hitting the runway, which was spared because the RAF would need it. The occupation of the Falklands ended on 15 June with the surrender by Argentinian forces. The Victors participation in the war had been crucial. Not only did the Victors tank for the Vulcans on the Black Buck series of missions but they also air refuelled Harriers from Britain to Ascension and after the surrender the tankers were kept fully occupied

Vickers Valiant B(K).1 WZ365 in flight.

Avro Vulcan B.2 XH558 in 1986. *Author*

supporting Nimrod and C-130 Hercules operations. In total the twenty-three Victors logged 1,980 hours 30 minutes (including 1,105 hours by thirteen Victors of No.55 Squadron) throughout the entire conflict.

As a result of experience gained in the Falklands, the Vulcan, like the Valiant and Victor before it, adopted the role of tanker in the last years of its career. Six Vulcan K.Mk.2 tanker conversions were hastily carried out at Woodford to meet the urgent need for extra tanker capacity and they served until March 1984, when there were sufficient VC-10s to take over the tanker role. On 30 March 1984 No.50 Squadron disbanded, thus marking the end of the Vulcan in RAF service. Victor deployment support to the South Atlantic continued until Mount Pleasant Airport opened in May 1985 and Wing Commander Martin Todd, OC No.55 Squadron flew the last Victor out of Ascension and back to Marham on 10 June. No.101 Squadron, with its five VC 10 K.2 and four VC 10 K.3s became operational in 1984 and added a significant improvement in Air-to-Air Refuelling (AAR) capability. The requirement for a strategic tanker to meet the commitment of the kind demonstrated in the South Atlantic led to an order for six TriStar tankers with an additional three authorised for conversion in 1986. At Marham the 'Tanker Trash' soldiered on in front line service fulfilling day to day air refuelling operations for Strike Command's Jaguars and Phantoms while all around them the station moved vigorously into the Tornado age. In September 1983 three No.617 Squadron Tornados visited Canada, flying the 3,500-mile journey to the Toronto International Airport with the aid of Victor and Vulcan tankers. A total of twenty-four K 2 tankers equipped Nos 55 and 57 Squadrons at RAF Marham until No.57 Squadron disbanded on 30 June 1986, leaving No.55 Squadron as the sole operator of the Victor.

In October 1984 the RAF returned to the United States after a gap of 4 years to take part in the annual US Strategic Air Command

Bombing Competition, known as Giant Voice. The RAF first entered Giant Voice in 1951 but in 30 years recorded success only in 1974 when one Vulcan crew won the Mathis Trophy and another won a navigation trophy. No.617 Squadron's six Tornados needed tanker support provided by No.55 Squadron's Victors to compete in Operation Prairie vortex at Ellsworth Air Force Base, South Dakota against B-52s of Strategic Air Command and F-111s of the Tactical Air Command. F-111Cs of the Royal Australian Air Force also took part. In total forty-two crews were competing for three trophies. The competition involved two phases of bombing sorties over the low-level ranges in Montana, Nevada, South Dakota and Wyoming, and extended over eight weeks. The first six weeks of the detachment being spent as a work up period to allow everyone to acclimatize, settle in and sort out the aircraft before the competition proper began. The first phase comprised a single 5?-hour daylight 'hi-lo-hi' bombing mission in which terrain following and ECM were employed to avoid detection and attacks by interceptors and simulated Strategic Air Missiles (SAMs); live practice bombs were dropped on invisible targets using offset blind bombing techniques. A low-level dash and high-level cruise return completed the mission. Phase 2 involved a 6-hour mission and was flown over two separate courses, one in daylight and the other at night. Multiple targets were attacked at high and low level using tone bombs while evading multiple threats from interceptors and missiles. Throughout the missions timing was recorded to within one-second accuracy as was navigation and bombing. It must be remembered that the Tornado was the only aircraft in the competition that demanded in-flight refuelling, by Victor K.2 tankers of No.55 Squadron, requiring split-second timing to avoid acquiring penalty points. The Tornados required at least two AAR brackets per sortie and in total 111 day and night join-ups were successfully made. The Le May and Meyer trophies had never been out of the USA before but No.617 Squadron won both of these trophies and the Mathis was missed only because the radar failed in one of the Tornados at a crucial point in the competition.

In October 1985 six Tornados of No.27 Squadron together with Victor tankers of No.57 Squadron, competed in that year's Bombing Competition at Ellsworth. **Squadron Leader Terry Cook** recalls: 'The early sorties were designed to be mini replicas of the competition missions with both Victors and Tornados flying for about 2? hours in preparation for the 6? hour competition routes. The whole RAF detachment was supporting two teams in the competition with two Tornado crews and two Victor crews in each team. Twice on each competition sortie the Tornados had to refuel from the Victors. Without the benefit of British Military Air Traffic or Fighter Controllers the airborne rendezvous had to be carried out by either aircraft type or, more usually, a combination of both. Once in company, the Victor crews assumed responsibility for navigation and timing in order to allow the Tornado crews a short period of relaxation. Without the benefit of modern, electronic navigation aids the Victors always took the Tornados to the end of the refuelling leg (approximately 200 nautical miles) within a few seconds of the required time! There is no doubt that had the Victors been allowed to enter the navigation parts of the competition they would have achieved very creditable results against the much better equipped USAF KC-135 tankers. On many sorties the Victors also acted in other roles, especially in negotiating extra refuelling airspace to meet and refuel Tornados that were short of fuel. One Victor crew even offered to fly along the low level route at medium level and to act as a relay between the Tornado, which had a poor radio, and the various range controllers.

With just three competition sorties per crew we all worked hard to have all the aircraft serviceable and to launch two Tornados and four Victors on each competition day. Generally all the aircraft were available to fly

although the Tornado crews usually had a preference as some aircraft dropped better scoring bombs than others. For the final day of the competition, one Victor and two Tornados were not available due to a series of problems; indeed, one Tornado was at a base in Wyoming after suffering an engine failure. All six competition aircraft, however, launched on time for the two sorties.'

On 30 June 1986, No.57 Squadron disbanded. Its K.2s were handed over to No.55 Squadron, which would continue to operate the Victor tankers for seven more years after that. No one could possibly have foreseen that before then the Victors would still have one more war to fight. In August 1990 No.55 Squadron's Victors were supporting RAF Jaguars at the Reconnaissance Air Meet in Texas when the recall of all tankers to the UK was ordered. Within 24 hours the 'Tanker Trash' were back at Marham and within 48 hours they were operating over France and Sicily to deploy fast jets to the Persian Gulf where conflict had begun on 2 August when Iraq's president Saddam Hussein's armies invaded Kuwait. On 7 August President George Bush ordered Operation Desert Shield to liberate Kuwait. USAF Lieutenant General Charles A. Horner, the Allied coalition's supreme air commander, began co-ordinating all air actions related to the build-up and within days, established HQ Central Command Air Forces (Forward) in Saudi Arabia. Initial Air Force planning was largely concentrated on defending Saudi Arabia from invasion and the first priority therefore was the neutralisation of advancing Iraqi tank and troop columns. The coalition air forces faced 750 Iraqi combat aircraft, 200 support aircraft, Scud surface-to-surface missiles, chemical and biological weapon capability, 'state-of-the-art' air defences, ten types of surface-to-air missiles, around 9,000 anti-aircraft artillery pieces and thousands of small arms. The Iraqi air force had twenty-four main operating bases and thirty dispersal fields; many equipped with the latest in hardened aircraft

Avro Vulcan B.2 XH558 taking off in 1986. *Author*

shelters. As many as forty-five of the most important targets were situated in and around Baghdad, a city covering 254 square miles and one which was considered to be seven times more heavily defended than Hànôi had been in December 1972.

All the RAF's refuelling assets were needed to deliver the Tornado GR.1s, F.3s and Jaguars from Europe to Tabuk and Dhahran in Saudi Arabia and the former RAF Muharraq, now Bahrain International Airport. In London on 9 August the MoD had announced the forthcoming dispatch of a dozen each of Tornado F.3 air-defence fighters and Jaguar GR.1A attack aircraft. Operation Granby, the British contribution to Desert Shield/Desert storm had begun. By the time the Gulf War began at 23.40Z on Wednesday 16 January the Coalition had built an air force of 2,790 aircraft, over half of which were combat aircraft. Included in this total was the RAF contribution of 135 aircraft, which included forty-six Tornado GR.1/IA attack, and reconnaissance aircraft. The RAF had despatched almost its entire tanker fleet to the Gulf region. A maximum of nine VC 10 K.Mk.2/3s of No.101 Squadron and two TriStar K.Mk.1s of No.216 Squadron were at King Khalid Airport just outside Riyadh. On completion of their fighter-bomber positioning assignments one Victor arrived at Muharraq on 14 December for a formal hand-over from No.101 Squadron VC 10 detachment, which had been operating at the Saudi base for three months. The next day three more Victors arrived, followed shortly by two more crews, as **Wing Commander David Williams**, OC No.55 Squadron recalls: 'The initial requirement from RAF Strike Command was that the Victor detachment should support the Tornado F.3 and Jaguar missions only and the VC 10 detachment would support all Tornado GR.1 sorties. After all ground training was completed flying began on 16 December to air-refuelling "towlines" scattered throughout the Saudi Arabian airspace. It became appar-

ent that the rigid apportioning of receivers to tankers was impractical and the Victors supported all types of aircraft from the UK, Canada, France and the United States Navy and Air Force that were probe and drogue compatible. After two weeks of intensive flying, four crews returned to the UK so that the remaining four crews could be brought into the environment and be fully trained to a war footing by 12 January 1991. On this date additional Victor aircraft together with the two initial crews were positioned at Muharraq and by 16 January No.55 Squadron was a total of six aircraft, eight crews and ninety-nine ground crew. Further training sorties were flown until 16 January when at 2250 Zulu, two Victors led the first Muharraq Tornado GR.1 bombing mission into Iraq. The sortie was flown along "olive low trail", which was a track south of the Iraq border but concluded with a northerly heading to cast off the receivers into the heart of Iraq. Olive trail then became the bread and butter for the rest of the war. To meet all contingencies and to ensure that fuel was available for the Tornados on their return from the mission, all Victor aircraft were refuelled to the maximum 123,000lb for take-off. The early sorties were affected by very poor weather along the refuelling tracks and consequently, the aircraft consumed treble the normal fatigue. As experience grew, the take-off fuel was adjusted and the fatigue penalty was reduced. On 19 January an additional Victor was sent to supplement the other six. Up to a maximum of fourteen sorties were flown over the Persian Gulf in support of attack missions and air defence patrols and together with 138 olive trails and numerous other patrols, 299 sorties were flown over the 42 day war, an average of thirty-three missions per crew. The Victor detachment achieved every objective and did not fall down on any operational sortie. It was tight at times and the need for flexibility, excellent engineering support and good airmanship saved the day and produced a 100 per cent success rate.'

Avro Vulcan B.2 taking off. *Avro*

The end of the war came suddenly and unexpectedly when a cease-fire was declared on 28 February 1991. Following the expiry of the United Nations ultimatum for Iraq to withdraw from Kuwait by 15 January, air operations had started on 17 January and had continued for a total of 42 days, while the ground war, starting on 24 February, had lasted only for 100 hours. However, these operations resulted in the complete defeat of the Iraqi forces.

During Granby and Desert Storm ten of No.55 Squadron's Victor K.2s, the oldest aircraft in the campaign, flew 299 sorties and by 18 March 1991 all had returned to Marham. The RAF retained the Vulcan for display purposes, one of which was XH558, which continued with the Vulcan Display Flight until 20 September 1992. The Victors of No.55 squadron flew busily right up to the day of their retirement on 30 September 1993. On 15 October No.55 Squadron was disbanded but for delivery purposes a few Victors even flew after this date. The last Victor flight to be made was on 30 November 1993 when XH672 *Maid Marion* was flown to Shawbury to be dismantled for transportation to the RAF Museum at Cosford. The first Victor was delivered to the Royal Air Force on 28 November 1957 and the last flight of a Victor was over 36 years from first delivery.

It was entirely due to British ingenuity, design and engine development that RAF Bomber Command Wellingtons, Stirlings, Lancasters and Halifaxes decisively won the European air war 1939–45 while a generation later the RAF V-force played a very significant role in the Cold War deterrent strategy. Never again will this nation conceive, design and build not one but four and then three entirely British manufactured military aircraft powered by British-made engines and armed with British weaponry in any one generation.

V for Victory. V for V-Force.

Victor tanker XH650 of No.55 Squadron air refuelling two Lightning F.3s of No.29 Squadron. *MoD*

Victor tanker XA933 with a No.29 Squadron Lightning F.3 during a refuelling exercise. *Dick Bell*

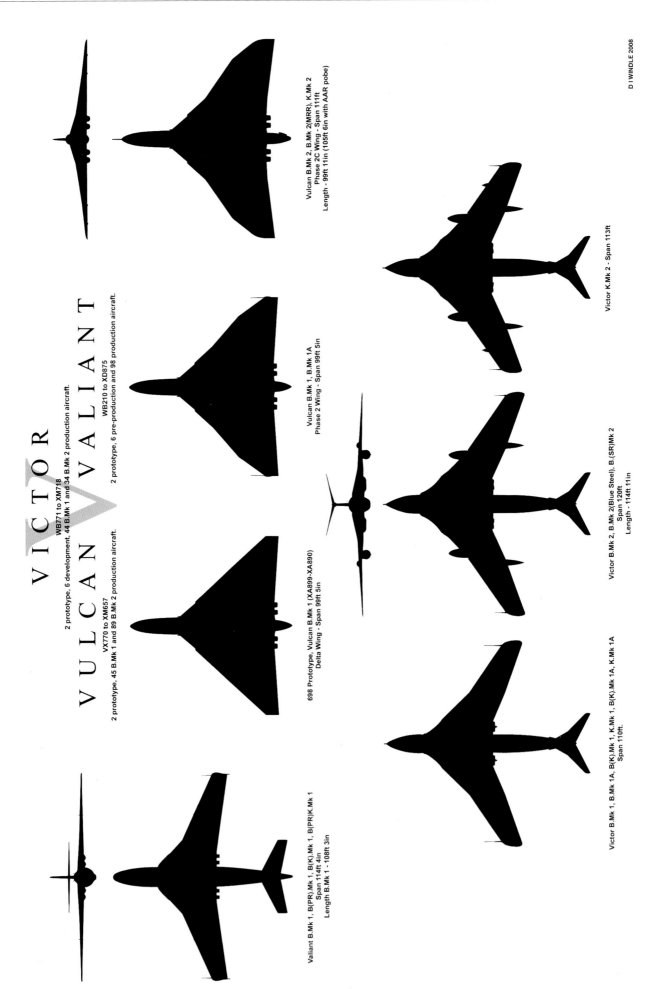

VICTOR
WB771 to XM718
2 prototype, 6 development, 44 B.Mk 1 and 34 B.Mk 2 production aircraft.

VULCAN
VX770 to XM657
2 prototype, 45 B.Mk 1 and 89 B.Mk 2 production aircraft.

VALIANT
WB210 to XD875
2 prototype, 6 pre-production and 98 production aircraft.

Vulcan B.Mk 2, B.Mk 2(MRR), K.Mk 2
Phase 2C Wing - Span 111ft
Length - 99ft 11in (105ft 6in with AAR pobe)

Vulcan B.Mk 1, B.Mk 1A,
Phase 2 Wing - Span 99ft 5in

698 Prototype, Vulcan B.Mk 1 (XA899-XA890)
Delta Wing - Span 99ft 5in

Valiant B.Mk 1, B(PR).Mk 1, B(K).Mk 1, B(PR)K.Mk 1
Span 114ft 4in
Length B.Mk 1 - 108ft 3in

Victor K.Mk 2 - Span 113ft

Victor B.Mk 2, B.Mk 2(Blue Steel), B.(SR)Mk 2
Span 120ft
Length - 114ft 11in

Victor B.Mk 1, B.Mk 1A, B(K).Mk 1, K.Mk 1, B(K).Mk 1A, K.Mk 1A
Span 110ft.

D I WINDLE 2008

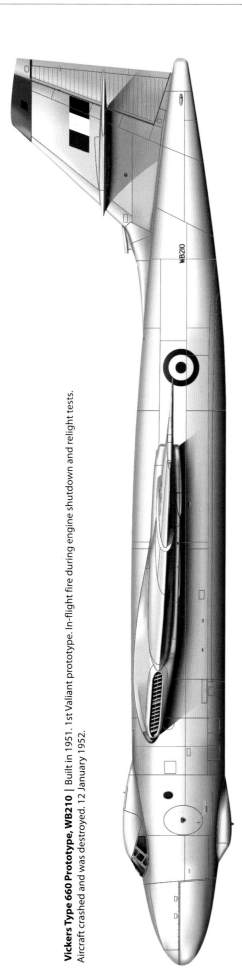

Vickers Type 660 Prototype, WB210 | Built in 1951. 1st Valiant prototype. In-flight fire during engine shutdown and relight tests. Aircraft crashed and was destroyed. 12 January 1952.

Valiant B(PR).Mk 1, WP223, No.18 Squadron | Built in 1955. 19th production Valiant. Scrapped 1965.

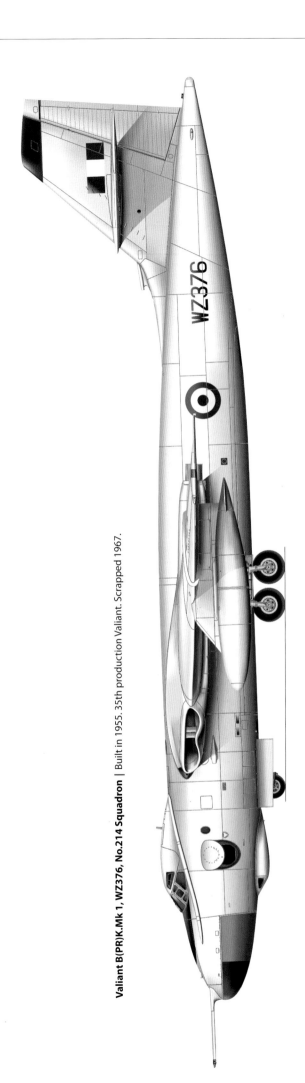

Valiant B.Mk 1, WZ365 | Built in 1955. 24th production Valiant. Scrapped 1965.

Valiant B(PR)K.Mk 1, WZ376, No.214 Squadron | Built in 1955. 35th production Valiant. Scrapped 1967.

Valiant B(K).Mk 1, WZ390, No.214 Squadron | Built in 1956. 45th production Valiant. Scrapped 1965.

Valiant B(PR)K.Mk 1, WZ392, No.543 Squadron | Built in 1956. 47th production Valiant. Scrapped 1965.

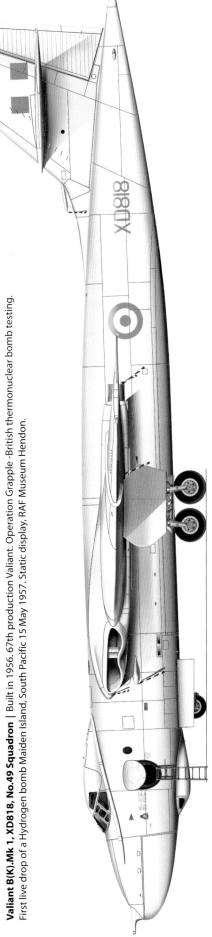

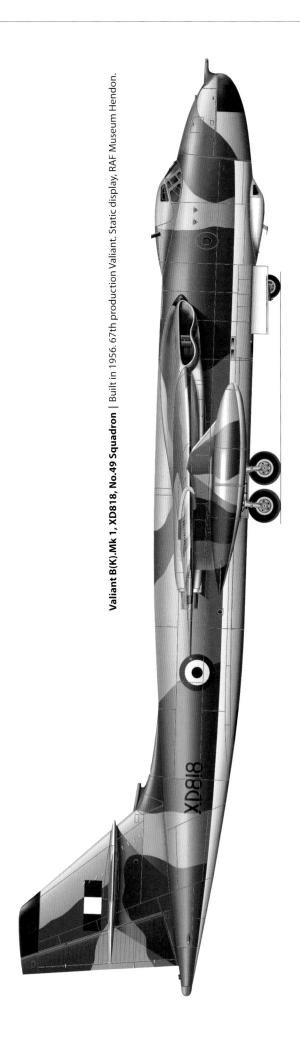

Valiant B(K).Mk 1, XD818, No.49 Squadron | Built in 1956. 67th production Valiant. Operation Grapple -British thermonuclear bomb testing. First live drop of a Hydrogen bomb Maiden Island, South Pacific 15 May 1957. Static display, RAF Museum Hendon.

Valiant B(K).Mk 1, XD818, No.49 Squadron | Built in 1956. 67th production Valiant. Static display, RAF Museum Hendon.

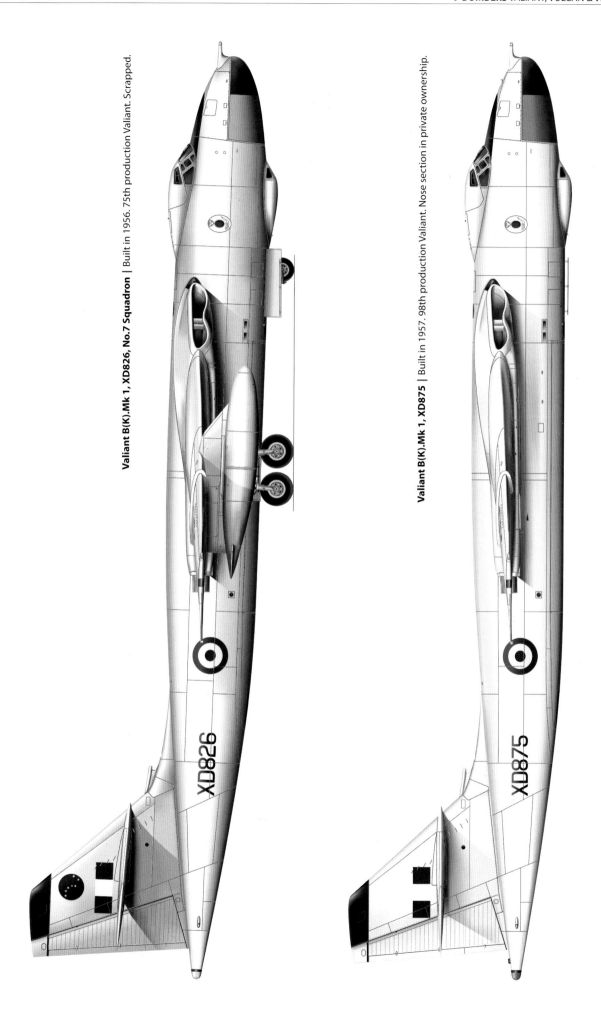

Valiant B(K).Mk 1, XD826, No.7 Squadron | Built in 1956. 75th production Valiant. Scrapped.

Valiant B(K).Mk 1, XD875 | Built in 1957. 98th production Valiant. Nose section in private ownership.

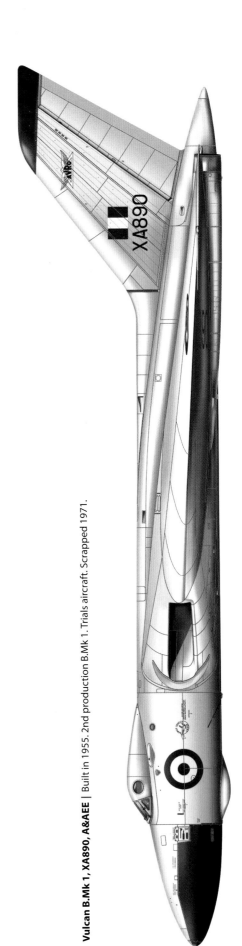

Avro Type 698 Prototype, VX770, A&AEE | Built in 1952. 1st Vulcan prototype. Structural failure of the starboard wing during flying display. Aircraft crashed and was destroyed. RAF Syerston 20 September 1958.

Vulcan B.Mk 1, XA890, A&AEE | Built in 1955. 2nd production B.Mk 1. Trials aircraft. Scrapped 1971.

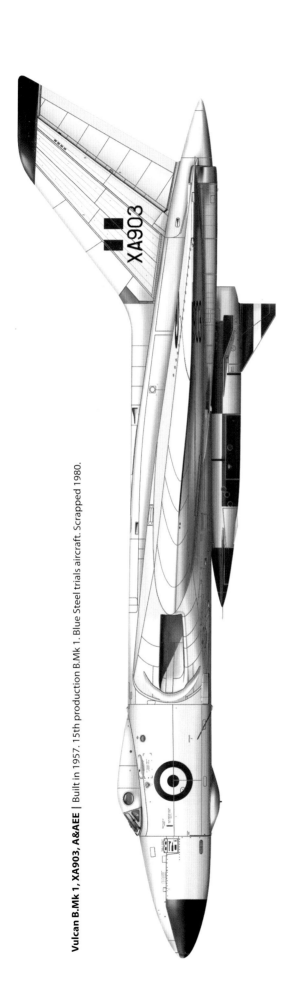

Vulcan B.Mk 1, XA903, A&AEE | Built in 1957. 15th production B.Mk 1. Blue Steel trials aircraft. Scrapped 1980.

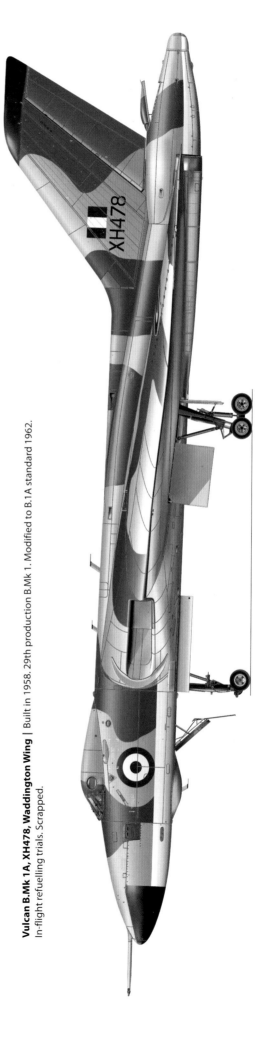

Vulcan B.Mk 1A, XH478, Waddington Wing | Built in 1958. 29th production B.Mk 1. Modified to B.1A standard 1962. In-flight refuelling trials. Scrapped.

Vulcan B.Mk 2, XH533, A&AEE | Built in 1958. 1st production B.Mk 2. (before ECM tail modification). One of the first 10 production B.Mk 2s to retain the smaller B.Mk 1 engine air intakes. Scrapped 1970.

Vulcan B.Mk 2, XH558 | Built in 1960. 12th production B.Mk 2. Last Vulcan in RAF service. Bruntingthorpe 23 March 1993. Restored to flying condition.

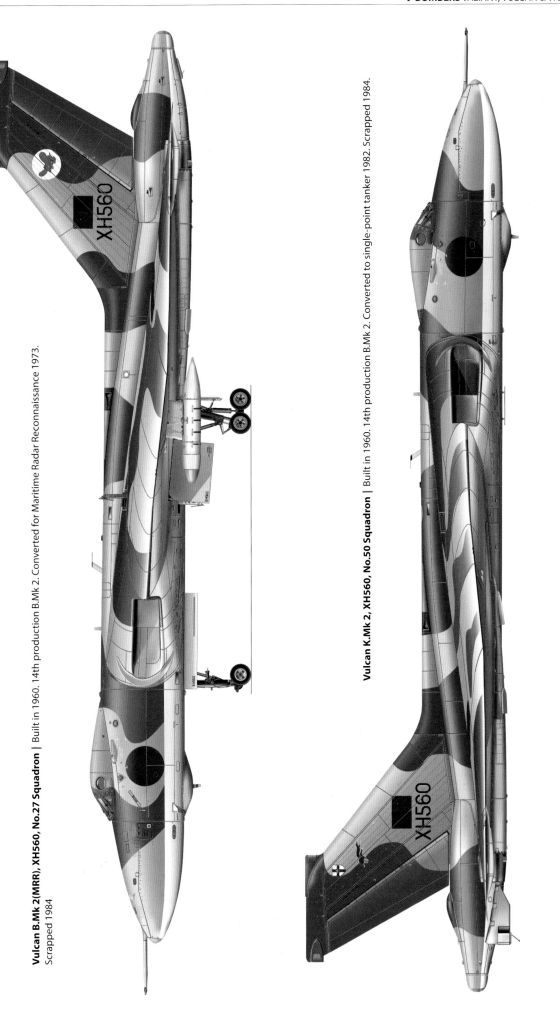

Vulcan B.Mk 2(MRR), XH560, No.27 Squadron | Built in 1960. 14th production B.Mk 2. Converted for Maritime Radar Reconnaissance 1973. Scrapped 1984

Vulcan K.Mk 2, XH560, No.50 Squadron | Built in 1960. 14th production B.Mk 2. Converted to single-point tanker 1982. Scrapped 1984.

Vulcan B.Mk 2, XL361, No.617 Squadron | Built in 1962. 33rd production B.Mk 2. Modified for Blue Steel stand-off missile. Static display at Goose Bay, Canada.

Vulcan B.Mk 2, XL443, No.35 Squadron | Built in 1962. 46th production B.Mk 2. Modified for Blue Steel stand-off missile. Scrapped 1982.

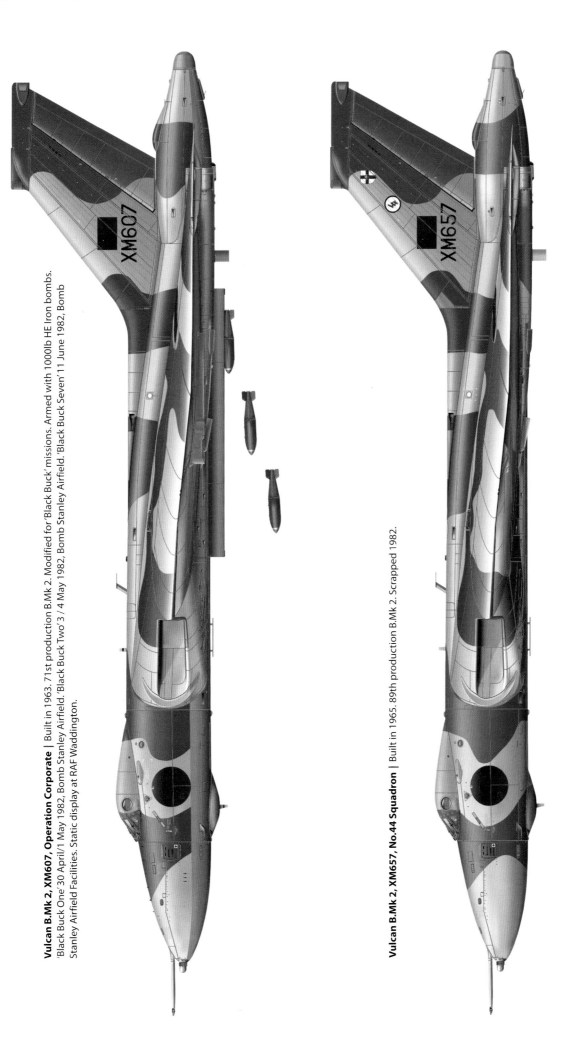

Vulcan B.Mk 2, XM607, Operation Corporate | Built in 1963. 71st production B.Mk 2. Modified for 'Black Buck' missions. Armed with 1000lb HE Iron bombs. 'Black Buck One' 30 April/1 May 1982, Bomb Stanley Airfield. 'Black Buck Two' 3 / 4 May 1982, Bomb Stanley Airfield. 'Black Buck Seven' 11 June 1982, Bomb Stanley Airfield Facilities. Static display at RAF Waddington.

Vulcan B.Mk 2, XM657, No.44 Squadron | Built in 1965. 89th production B.Mk 2. Scrapped 1982.

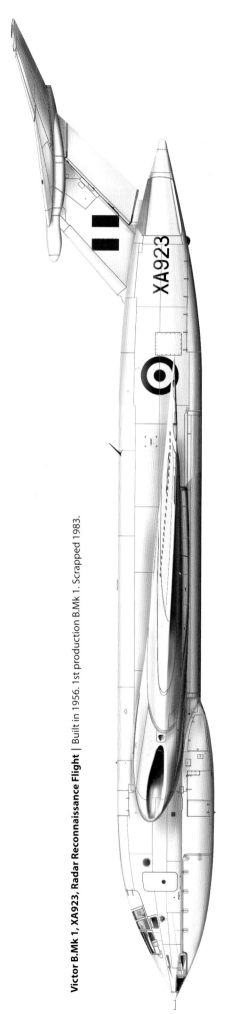

Handley Page H.P. 80 Prototype, WB771, A&AEE | Built in 1952. 1st Victor prototype. Loss of tail-plane during low-level test flight. Aircraft crashed and was destroyed. Cranfield 14 July 1954.

Victor B.Mk 1, XA923, Radar Reconnaissance Flight | Built in 1956. 1st production B.Mk 1. Scrapped 1983.

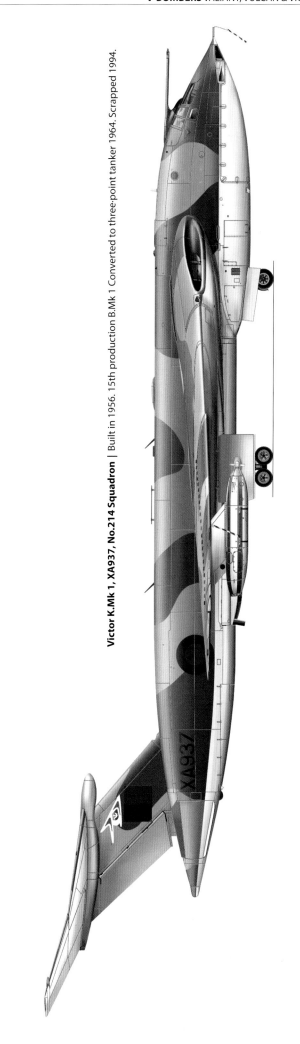

Victor B(K).Mk 1, XA937, No.214 Squadron | Built in 1956. 15th production B.Mk 1 Converted to three-point tanker 1964. Later re-designated Victor K.Mk 1. Scrapped 1994

Victor K.Mk 1, XA937, No.214 Squadron | Built in 1956. 15th production B.Mk 1 Converted to three-point tanker 1964. Scrapped 1994.

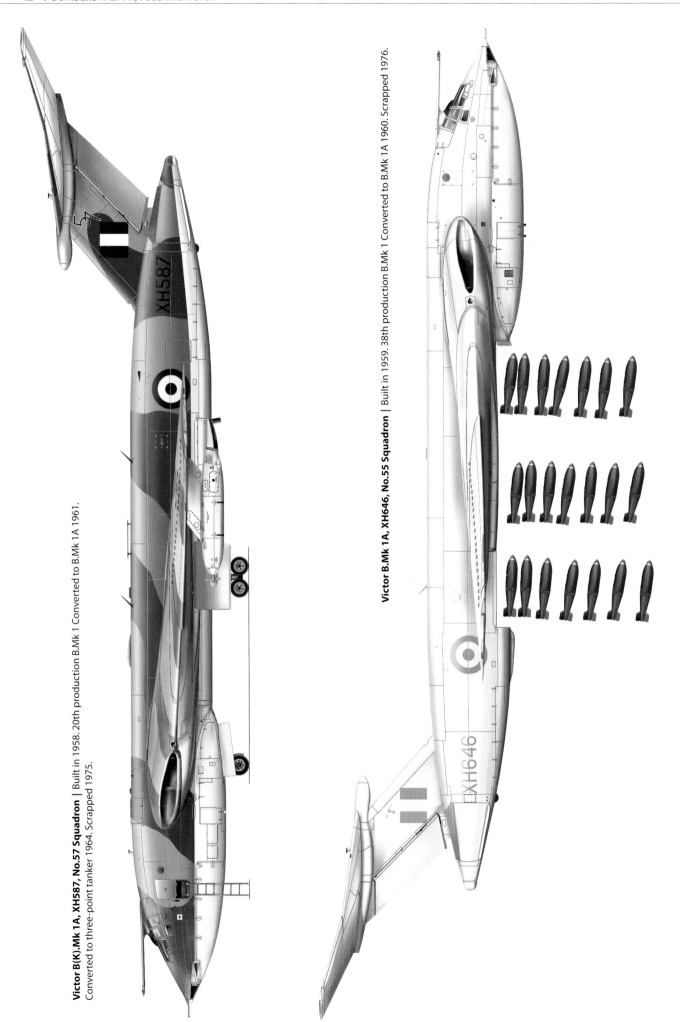

Victor B(K).Mk 1A, XH587, No.57 Squadron | Built in 1958. 20th production B.Mk 1 Converted to B.Mk 1A 1961. Converted to three-point tanker 1964. Scrapped 1975.

Victor B.Mk 1A, XH646, No.55 Squadron | Built in 1959. 38th production B.Mk 1 Converted to B.Mk 1A 1960. Scrapped 1976.

Victor B.Mk 1A(K2P), XH648, No.57 Squadron | Built in 1959. 40th production B.Mk 1 Converted to B.Mk 1A 1960. Converted to two-point tanker 1965. Static display at the IWM Duxford.

Victor B(SR).Mk 2, XL230, No.543 Squadron | Built in 1961. 23rd production B.Mk 2 Converted to the Strategic Reconnaissance role 1964. Scrapped 1973.

Victor K.Mk 2, XL231, No.55 Squadron | Built in 1961. 24th production B.Mk 2 Converted to three-point tanker 1970 Desert Storm 1991, 'Lusty Lindy' 16 missions. Taxiable condition, Elvington.

Victor B.Mk 2, XM715, No.100 Squadron | Built in 1962. 31st production B.Mk 2 Taxiable condition, Bruntingthorpe.

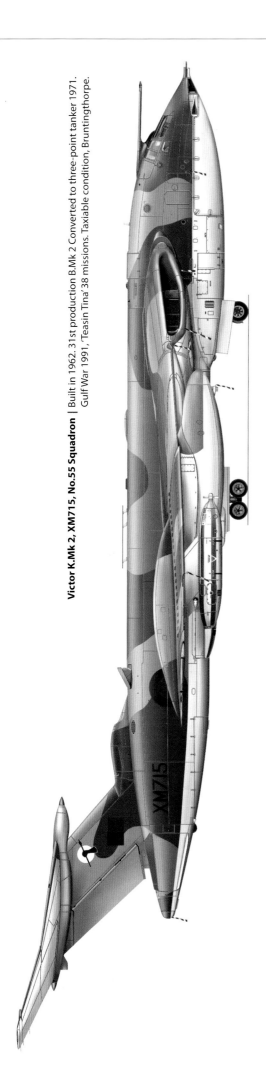

Victor K.Mk 2, XM715, No.55 Squadron | Built in 1962. 31st production B.Mk 2 Converted to three-point tanker 1971. Gulf War 1991, 'Teasin Tina' 38 missions. Taxiable condition, Bruntingthorpe.

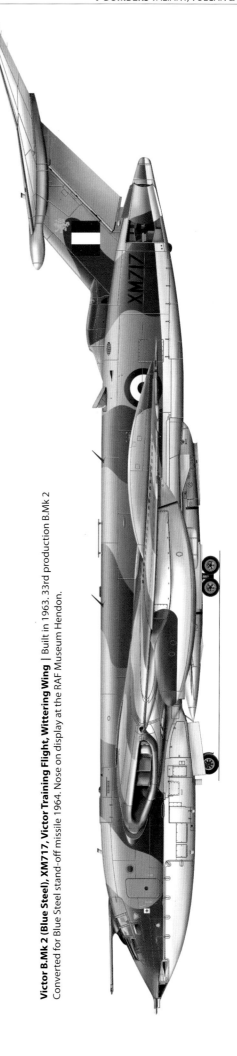

Victor B.Mk 2 (Blue Steel), XM717, Victor Training Flight, Wittering Wing | Built in 1963. 33rd production B.Mk 2 Converted for Blue Steel stand-off missile 1964. Nose on display at the RAF Museum Hendon.

■ V-BOMBER PRODUCTION AND SERIAL BLOCKS

VALIANT

Type 660	prototype WB210	1
Type 667	WB215	1
Valiant B.1	WP199-WP203 WP204, WP206-WP216, WP218, WP220, WP222, WZ361-WZ375, WZ377	36
Valiant B (PR) 1	WP203, WP217, WP219, WP221 WP223, WZ376, WZ378, WZ379, WZ381, WZ383, WZ384	11
Valiant B (PR) K.1	WZ380, WZ382, WZ389-WZ399	13
Valiant B (K).1	WZ400-WZ405, XD812-XD830, XD857-XD875	44
Valiant B (K).1	XD867-XD893, XE294-XE299	(24) cancelled
Valiant Mk.2	prototype WJ954	1
Total		**107**

VULCAN

Type 698	prototypes VX770, VX777	2
Vulcan B.1 and B.1A	XA889-XA913, XH475-XH483, XH497-XH506, XH532	45
Vulcan B.2	XH533-XH539, XH554-XH563, XJ780-XJ784, XJ823-XJ825, XL317-XL321, XL359-XL361, XL384, XL392, XL425-XL427, XL443-XL446, XM569-XM576, XM594, XM595, XM596	
	(airframe used as a fatigue test specimen), XM597-XM612, XM645-XM657	89
Total		**136**

VICTOR

Type HP80	prototypes WB771, WB712	2
Victor B.1 and B.1A	XA917-XA941, XH587-XH594, XH613-XH621, XH645-XH651, XH667	50
Victor B.2	XH668-XH615, XL158-XL165, XL188-XL193, XL230-XL233, XL511-XL513, XM714-XM718,	34
	XM719-XM721, XM745-XM756, XM785-XM794	(25) cancelled
Total		**86**

■ V-BOMBER UNITS

Unit	Station	Dates
VALIANT		
232 OCU	Gaydon	June 1955, 1st Course becoming No.138 Squadron at Wittering 6 July 1955
7 Squadron	Honington/Wittering	1 November 1956 Valiant B(PR).1. 1957 B.l/B(K).l. August 1961, B(PR)K.l. Disbanded 30 September 1962
18 Squadron	Finningley	17 December 1958-31 March 1963 (Valiant B.l in a joint ECM role)
49 Squadron	Wittering/Marham	1 May 1956 (Valiant B.l). Disbanded 1 May 1965
90 Squadron	Honington	1 January 1957 (Valiant B(K).l/B(PR).l/B(PR)K.l). Disbanded 1 March 1965
138 Squadron	Wittering	February 1955. Disbanded 1 April 1962
148 Squadron	Marham	1 July 1956. Disbanded 1 May 1965
199 Squadron	Hemswell/Honington	May 1957 – 17 December 1958 (renumbered as No.18 Squadron at Finningley)
207 Squadron	Marham	June 1956. Disbanded 1 May 1965
214 Squadron	Marham	21 January 1956. Disbanded 1 March 1965
543 Squadron	Wyton	Second squadron to be formed from 232 OCU. To Wyton 18 November 1955. Valiants withdrawn, December 1964
VULCAN		
230 OCU	Waddington	August 1956. Vulcan B.l/1A/2s from January 1957. Disbanded 1980
9 Squadron	Coningsby/Cottesmore Akrotiri/Waddington	1 March 1962 – disbanded as a Vulcan operator, 1 May 1982
12 Squadron	Coningsby/Cottesmore	1 July 1962. Disbanded 31 December 1967
27 Squadron	Scampton	1 April 1961. Disbanded 29 March 1972. Re-formed at Scampton, 1 November 1973. Operated Vulcan B.2(MRR) aircraft before being disbanded again 31 March 1982
35 Squadron	Coningsby/Akrotiri/Scampton	1 December 1962. Disbanded 1 March 1982
44 (Rhodesia) Sqn	Waddington	10 August 1960. Disbanded 21 December 1982
50 Squadron	Waddington	1 August 1961. Disbanded 31 March 1984
83 Squadron	Waddington	21 May 1957. Disbanded 31 August 1969
101 Squadron	Finningley/Waddington	15 October 1957. Disbanded 4 August 1982
617 Squadron	Scampton	1 May 1958. Disbanded 31 December 1981
VICTOR		
232 OCU	Gaydon/Cottesmore	28 November 1957 – 1961
10 Squadron	Cottesmore	15 April 1958. Disbanded 1 March 1964
15 Squadron	Cottesmore	15 April 1958. Disbanded 1 October 1964
55 Squadron	Honington/Marham	1 September 1960. Disbanded 15 October 1993
57 Squadron	Honington/Marham	1 January 1959 – 30 June 1986
100 Squadron	Wittering	1 May 1962 – 30 September 1968
139 Squadron	Wittering	1 February 1962. Disbanded 31 December 1968
214 Squadron	Marham	1 July 1966. Disbanded 28 January 1977
543 Squadron	Wyton	May 1965. Disbanded 24 May 1974

Modelling the V-Bombers

Presumably due to the aircrafts' size, plastic injection-moulded construction kits of the three V-Bombers appear to be restricted to the smaller scales – at the time of writing that is – there being none currently available above 1/72 scale, although a limited run vacform kit in 1/48 scale was produced in the 1990s but is sadly no longer in production.

DRAGON/CYBERHOBBY   1/200 scale

Avro Vulcan B.2

Starting with the smallest, Dragon/Cyber-Hobby initially produced an injection-moulded plastic 'Cold War' Vulcan B.2 armed with a Blue Steel nuclear missile. Moulded in grey plastic, with clear plastic canopy, the upper and under surfaces of the wings/fuselage were moulded in halves.

More recently, the kit has been re-released in the 'Modern Air Power Series' and now comprises 70 plus parts with newly tooled one piece wings with detailed engine intakes and exhaust nozzles. Two ABM-45A Shrike missiles and an ALQ-101 electronic counter measures pod are included. As with the original release, the undercarriage can be assembled in either an up or down posi-

tion. The decal sheet, printed by Cartograf, features markings for two aircraft from Nos.27 and 44 Squadrons, based on Ascension Island in 1982 during the Falklands War.

Dragon/CyberHobby Vulcan B.2 box top re-released in the 'Modern Air Power Series' and now comprises 70+ parts with newly tooled one-piece wings with detailed engine intakes and exhaust nozzles.

PIT ROAD (ALSO GREAT WALL HOBBY)   1/144 scale

The growing popularity of this scale means that an increasing range of subjects are now becoming available, not least of which are all three V-Bombers. Given the size of these aircraft the scale is ideal and allows for plenty of detail to be included.

Avro Vulcan B.2 (Falklands) and K.2

Probably the best currently available 1/144 scale model of the Vulcan is their B.2 which is a modern injection-moulded plastic tooling with excellent surface and engraved panel line detail. The complete wing/fuselage is formed from upper and lower halves to which are added all the details including the intakes, which are good and deep.

The initial release depicted a pair of 'Falklands War' aircraft, XM597 and XM607, complete with Shrike missiles and ALQ-101 electronic counter-measures pod. The identical model has also been released recently by Great Wall Hobby in their own boxing.

These mouldings have also been released as a Vulcan K.2 tanker, under both the Pit Road and Great Wall Hobby labels. The same selection of sprues is included as the B.2 boxings, with an extra sprue for the hose and drogue unit and new decals, depicting No.50 Squadron aircraft XJ825 and XM571.

Handley Page Victor B.2

The Great Wall Hobby 1/144 scale Victor B.2 kit (announced in late 2014) has been released and there is every reason to hope that it will be just as good as their Vulcan.

Probably the best currently available 1/144 scale model of the Vulcan is the Pit Road (Great Wall Hobby) B.2.

The same mouldings have also been released as a Vulcan K.2 tanker, under both the Pit Road and Great Wall Hobby labels.

Box top of the Great Wall Hobby 1/144 scale Victor B.2.

CROWN/MINICRAFT
1/144 scale

Valiant B.1, Vulcan B.2, Victor B.2 and K.2
Crown/Minicraft produced a range of 1/144 scale V-Bombers including a Valiant B.1, Vulcan B.2, Victor B.2 and Victor K.2, but their current status is unknown.

ANIGRAND
1/144 scale

Anigrand led the 1/144 scale field for some time with their resin Vulcan B.2.

Anigrand have also produced both the Victor B.2 and Victor K.2 in their 1/144 scale resin range.

To complete the V-Bomber trio, Anigrand also produce a Valiant B.1 which is another competent and acceptable resin kit.

Avro Vulcan B.2
Anigrand led the 1/144 scale field for some time with a resin Vulcan B.2. Typical of the company's products, it is cast in cream resin with acceptable detail. The kit comes with what is presumably supposed to represent a Blue Steel nuclear missile, but in reality seems to depict the unbuilt Avro Z101 proposal for a hypersonic research aircraft, which was essentially a Blue Steel airframe with a cockpit, nosewheel and two aft-mounted skids.

Markings are for an overall 'anti-flash' white, Blue Steel-armed aircraft XH554, or XM607 in the later Medium Sea Grey/Dark Green upper surface camouflage scheme. As is traditional with Anigrand the package also includes 'bonus' models, in this case an Avro 707 'scale model' Vulcan precursor, plus a DH 108 and the SRA.1 jet fighter flying boat.

Handley Page Victor B.2 and K.2
Anigrand have also produced both the Victor B.2 and Victor K.2. Again a Blue Steel nuclear missile is offered, for the Victor B.2, but this time it is a 'proper' production missile. Markings for an overall 'anti-flash' white Blue Steel-armed aircraft, XL158, and a grey/green (upper surfaces) camouflaged B.2, XL165. In this instance, the 'bonus' models are a Belvedere helicopter, a Blackburn Firecrest and a GAL-61 experimental flying wing glider!

The Victor K.2 offers markings for two Desert Storm aircraft, XM715 'Teasin Tina' and XL231 'Lusty Lindy', both in the Hemp/Grey scheme. Bonus kits this time are the Fairey FD.1, a Wessex HC.2 and an AW.52 Flying Wing bomber.

Vickers Valiant B.1
Anigrand also produce a Valiant B.1. Another competent and acceptable resin kit, with markings this time for a natural metal finished machine, XD814, and an overall 'anti-flash' white XD818. This time the 'bonus' kits are three more British experimental types, the Fairey FD.1 again, the low-speed HP115 and the Bristol 188.

WOLFPACK
1/144 scale

Avro Vulcan B.2
The Vulcan B.2 was also briefly available from Wolfpack. It came as virtually a one-piece casting that featured full bomb-bay detail. It also featured rub-down decals which were well printed but awkward to use.

WELSH MODELS

1/144 scale

Avro Vulcan B.2

Welsh Models market a mixed-medium (i.e. vacform/resin/white metal) Vulcan B.2 with markings for XL321 of No.617 Squadron, or in a separate boxing, XM570 in overall 'anti-flash' white. They also offer a mixed-medium Valiant B.1, available in several boxings, including overall 'anti-flash' white and grey/green camouflaged variants, one of which is a B(PR).1 reconnaissance variant.

These models feature a vacform fuselage, resin wings and flying surfaces and white metal undercarriage detail. The decals are good too.

Welsh Models market a mixed-medium (ie vacform/resin/white metal) Vulcan B.2.

LINDBERG

1/96 scale

Vulcan and Victor

Lindberg released a 1/96 scale Vulcan and Victor in the mid-1950s. As expected from kits of this vintage they were 'of their time', with raised panel lines and were very basic with not a lot of detail, although the rudder, ailerons and elevators were all designed to move. They also featured removable panels. The Vulcan kit represented the second prototype, VX777, originally flown in 1953, the markings for which are on the decal sheet.

The Victor kit was also based on one of the two prototypes, WB775, again for which the decal sheet has the markings, rather than a production Victor B.1. It appears that the kit

Current box top presentation of the Lindberg 1/96 Vulcan.

is currently available under the Lindberg label (at the time of writing) in the guise of a camouflaged operational machine and comes with a refuelling truck and ground crew.

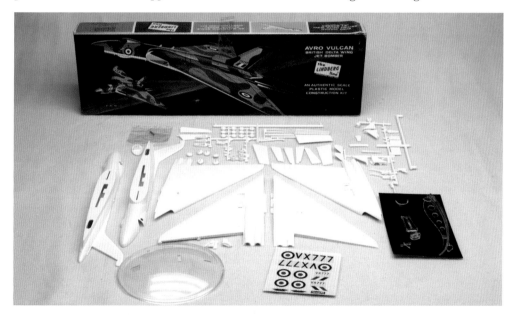

Left: The original box top presentation of the Lindberg 1/96 scale Vulcan kit.

Below left: The original 1/96 scale Lindberg Victor box top.

Below : The original 1/96 scale Lindberg Victor sprues.

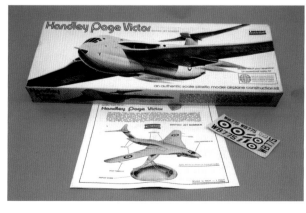

FROG

1/96 scale

Valiant, Vulcan and Victor

Frog produced all three V-Bombers (Valiant, Vulcan and Victor) to 1/96 scale in the 1950s. The Frog Vulcan and Victor were totally different to the Lindberg kits, the Vulcan having the later Phase 2 wing shape of the B.1 production version with the extended leading edges whereas the Lindberg Vulcan had the original prototype's 'triangular' wing. All the Frog V-bombers had engraved panel lines but are no longer currently available and are considered collector's items.

AIRFIX

1/72 scale

Avro Vulcan B.2

The Airfix Vulcan B.2 originates from the late-1970s and while now showing its age somewhat, is generally accurate. The airframe consists of a central fuselage split horizontally, with the outer wing panels in upper and lower halves, with a relatively sparse cockpit interior which is limited to a floor, two ejection seats, a decal instrument panel and two control columns. The crew access door is separate and can be modelled open and there is a boarding ladder.

The landing gear is very simple, but looks accurate enough, (see also Scale Aircraft Conversions set 72020 Airfix Vulcan undercarriage set), although the detail in the nose and mainwheel wells is non-existent. A Blue Steel nuclear missile is provided as is the modified bomb bay piece to accommodate the device.

The clear canopy parts are not too bad, but the major components are marred by large sink marks, especially in the engine intake areas and the engine exhaust pipes are virtually nothing but four holes in the back of the wings. (See Freightdog Models accessory sets 7201 and 7202 for accurate replacement parts). There are major seam gaps in virtually every area and the engine intakes, so prominent on the model, could have been moulded better. However, at the time of writing it is the only 1/72 scale Vulcan available and as can be seen from the accompanying photos, given a bit of care in construction, very acceptable models can be made from it.

Airfix 1/72 scale Vulcan box top and sprues

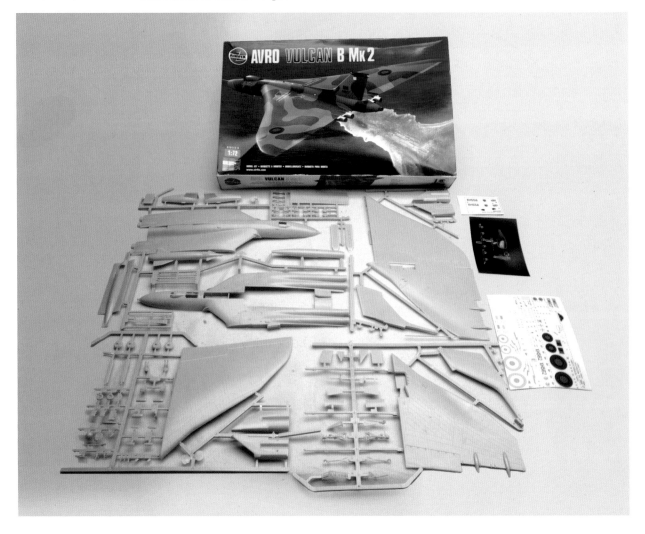

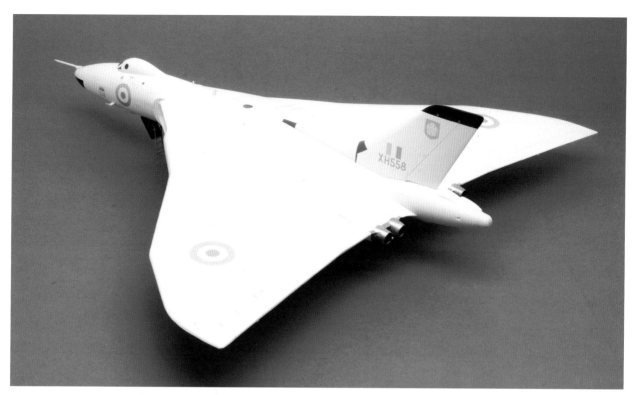

Vulcan B.2, XH558 of No.230 OCU based at RAF Finningley in 1963, made from the 1/72 scale Airfix kit.
Model by Paul Hughes

This page: Airfix 1/72 scale Vulcan B.2 finished as XH537 in the overall anti-flash white scheme with pale roundels, based at BAe Woodford, 1961, with Skybolt missiles from the Sharkit conversion set. *Model by Richard Farrar*

Opposite page: Vulcan B.2, XL320, based at RAF Scampton 1965, made from the 1/72 scale Airfix kit. *Model by Paul Hughes*

This page: Vulcan B.1, XA900 of 230 OCU based at RAF Waddington, 1957, made from the 1/72 scale Airfix kit using the DB Resin conversion set. *Model by Geoff Trenholme*

Opposite page:
Vulcan B.1A, XH504 based at RAF Waddington 1963, made from the 1/72 scale Airfix kit using the DB Resin conversion set. *Model by Geoff Trenholme*

Vulcan B.1a, XA895 of the Bomber Command Development Unit, made from the 1/72 scale Airfix kit using the DB Resin conversion set. *Model by Paul Hughes*

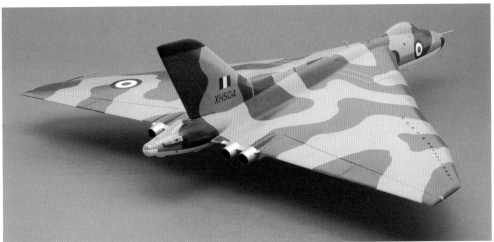

Vickers Valiant B.1

Released in 2013, the Airfix Valiant B.1 comes in a very large box – with surprisingly few parts and, as if to emphasise this, the wing contains only three main components: a large one-piece upper section and a pair of lower wing halves, albeit the ailerons are separate. Moulded in an easy to work with grey plastic, the fuselage halves may look flimsy but they have fairly thick walls and large locating pins so, when all the internal structure is in place, a very solid model results. Surface detail is minimal with engraved panel lines, but as the real aircraft was fully flush riveted it looks more than reasonable.

The cockpit interior is fairly basic with decals for the instrument panels and five nicely detailed seats for pilots and crew but only the pilots had ejection seats, the other members had to bale out a side door if in trouble. Also included are the control columns which extended from the side consoles and a ladder from the rear crew station up to the flight deck. Parts are included to model the door open or closed.

There are two sets of canopies – one without the upper cockpit windows and a round bomb aimer's aperture that only the first two prototypes were fitted with, although only WB215, the second prototype, can be modelled as WB210, had a completely different engine air intake arrangement for its Sapphire engines. Extended and closed in-flight refuelling probes are also included.

The bomb bay is well designed with the parts interlocking with the upper wing. There are two different bomb bay roofs, one for each of the two weapons load options –

the 'Blue Danube' nuclear bomb, or twenty-one 1,000lb conventional bombs. The bomb doors can be left opened or closed.

The mainwheel wells are well detailed large one-piece parts, which fit into tab slots very well. The nicely shaped air intakes and jet pipes have to be assembled with the wings as they slot into the fairing that then joins onto the wing. It might have been preferable to have added them at a later stage as once they have been painted they will need to be masked when the rest of the model is painted.

The mainwheel centres are separate from the tyres which makes painting them easier. Separate elevators are also provided but no separate flaps. The shoulder-mounted wing slots into the top of the fuselage, interlocking with the roof of the bomb bay making for a very solid join. After the tailplanes are assembled the elevators can be added then the finished assembly slotted into the aperture on the rudder. The vortex generators on the tailplanes should be on the under sides, but those on the side of the fin can be carefully cut off and replaced later after the fin flash decals have been applied.

The underwing fuel tanks have to be assembled as part of the underwing pylon and cannot be attached to the pylon at a later stage. If the underwing tanks are fitted, the undercarriage door needs fixing at a slight angle to clear the tanks, just as it was on the real aircraft. If the tanks are omitted then the door hangs vertically. The overall fit of the parts is excellent, there being no filler required.

Four aircraft are covered on the decal sheet, WB215, WZ404, XD823 and XD857.

Opposite page:
Valiant B.1, XD826, No 7 Squadron, RAF Honington, 1967, made from the 1/72 scale Airfix kit.
Model by Paul Hughes

Below: Valiant B.1, XD857 made from the 1/72 scale Airfix kit in the markings of No.49 Squadron based at RAF Marham during 1963 featuring low visibility national markings and the unit's red 'greyhound' emblem on the fin and large wing tanks.
Model by Tony O'Toole

Airfix Valiant Tanker Conversion

An innovative and welcome move was made by Airfix when they released a conversion set to upgrade their Valiant B.1 bomber kit, allowing it to be converted in to a B(K).1 tanker or a B(PR)K.1 strategic reconnaissance variant.

The conversion set comprises injection-moulded parts including a hose and drogue in-flight refuelling/tanker system built into the inside of the open bomb bay and a closed, one-piece, bomb bay door component for the photographic reconnaissance variant which has ten square camera ports

This page: Airfix 1/72 scale Valiant kit made in to a B(K).1 using the Airfix conversion set and finished as XD812 of No.214 Squadron, wearing full colour upper wing roundels and squared off styled serials. Although the majority of Valiants appear to have been kept in immaculate condition and the panel lines were not usually visible, some airframes were decidedly worn looking, especially the hard working air-to-air tankers, so XD812 has been finished in a weathered condition. All the vortex generators are etched brass from the Alley Cat conversion set.
Model by Tony O'Toole

Opposite page: Valiant B(K).1 made from the 1/72 scale Airfix kit, finished as XD875, the last Valiant built, as it looked while serving with No.138 Squadron, RAF Wittering, in 1961, in the hi-vis anti-flash white scheme.
Model by Richard Farrar

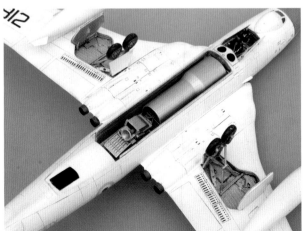

moulded into it and clear glazed parts. The conversion set also has a small decal sheet which provides additional serial and unit decals for one of each variant to be built. The tanker option is XD812 from No.214 Squadron in an overall anti-flash white scheme with the Nightjar bird from the official Squadron badge incorporated within the Air Refuelling Ltd company logo on the fin, while the reconnaissance option is WZ399 from No.543 Squadron in an overall High Speed Silver scheme with high visibility red trim and the Polar Bear insignia of the RCAF Winter Experimental Establishment (WEE) at Edmonton on the side of the nose, worn during winter trials in Canada.

MACH 2

1/72 scale

Vickers Valiant B.1

The French company Mach 2 was actually the first to offer a 1/72 scale version of the

Vickers Valiant, in the summer of 2006, in the form of a short-run injection-moulded kit. Typical of short run kits, it has thick sprues and seems quite crude, (although perhaps better than previous Mach 2 offerings), when compared to mainstream injection-moulded kits. It has about 60 parts in white plastic, plus some small transparent parts and a small decal sheet.

Panel lines are engraved (albeit not always straight!) and some re-scribing is needed. Optional parts include a representation of the 'Blue Danube' nuclear bomb, its transport trolley and two large external underwing fuel tanks.

Although the wing shape is reasonable, the shape of the forward fuselage does not look right and seems to be too short. The main cockpit area is simple with two seats and a floor – the rest of the cockpit, together with the remaining crew is not catered for.

Overall, assembly is not too difficult (for a short-run injection-moulded kit) as there are not that many parts, but there is no longer any need to buy the Mach 2 kit thanks to the Airfix offering.

REVELL (MATCHBOX)

1/72 scale

Handley Page Victor K.2

Matchbox initially released the HP Victor K.2 kit in 1/72 scale in the early 1980s. Moulded in their traditional multi-coloured plastic sprues, in this case white, grey and dark green, for its time it was a very good kit, the only drawback being the deep, trench-like, panel lines. Overall dimensions were acceptable, and there was a lot of plastic in the box!

With the demise of Matchbox, Revell took over the moulds, and re-released the kit in

the 2007. The kit is identical to the original Matchbox offering but is moulded in a single light grey plastic. The kit has decals for XL163 of No.57 Squadron and XL231 'Lusty Lindy' of No.55 Squadron, which is now preserved at the Yorkshire Air Museum.

Flightpath produce a Victor K.2 accessory set with resin intakes (DB2014), white metal cones for the tailplane and some photo etch (FHP72053) to enhance the IFR and cockpit framing with small rivets for the glazing.

Above: 1/72 scale Mach 2 Valiant box top and sprues

Right: Revell Victor box top and sprues.

Opposite page: Victor B.1, made from the 1/72 scale Matchbox kit with the FlightPath B.1 conversion set and finished as XH589, of No.15 Squadron, RAF Cottesmore, 1959. *Model by Richard Farrar*

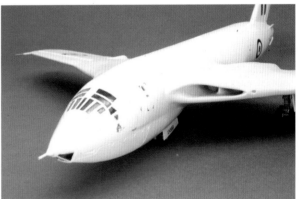

Victor B.2, made from
the 1/72 scale
Matchbox kit, XL513,
No 139 Squadron, RAF
Wittering, 1965, with
the FlightPath
conversion set.
Model by Richard Farrar

CONTRAIL   1/72 scale

Vickers Valiant B.1 and B.2

Contrail released two 1/72 scale vacform kits of the Valiant – the Valiant B.1 (early and revised issues) and the Valiant B.2, with the re-designed and strengthened airframe which was intended as a Pathfinder to mark targets for the main bomber force, of which there was just the one prototype made, WJ954, as the B.2 programme was abandoned by the RAF.

(WJ954 was used for tests for a few years, including testing the use of rockets to boost take-off, before being scrapped in 1958).

Although these vacforms are now quite old, they have stood the test of time well. Both kits included short run injection-moulded parts for the undercarriage legs. Aircraft in Miniature also reissued the Contrail B.1 and B.2.

RAREPLANES AND FORMAPLANE   1/72 scale

Both Rareplanes and Formaplane released vacform kits of the Vulcan B.2 and Victor B.2.

AEROCLUB   1/48 scale

Avro Vulcan B.2

Aeroclub produced a limited edition, multi-medium vacform kit in the late 1980s, consisting of four large sheets of thick vacformed plastic, one each for the upper and lower fuselage halves, one each for the port and starboard wings, plus another which contained the fin, the nose and main-wheel bays, fuselage formers and the cockpit interior. Two clear vacform canopies were

also included. The air intakes were moulded in resin, as were the jet pipes. A bag of white metal undercarriage parts was included, plus a sheet of decals, with instructions and a camouflage painting guide. The kit was limited to a production run of 100 it is believed, and sadly is no longer available, leaving the field wide open for some enterprising kit manufacturer to produce a Vulcan in 1/48 scale.

Aeroclub 1/48 scale vacform, resin and white metal Vulcan B.2, finished as XM575 in No.617 Squadron markings, when based at RAF Scampton in 1963.
Model by Richard Farrar

Another view of
Vulcan B.2 XM575.
Model by Richard Farrar

ACCESSORIES

RetroKit

RetroKit offer a 1/144 scale Vulcan K.2 resin conversion for the original Pit Road B.2 release. They also produce an open bomb bay conversion, again in resin, for the Vulcan B.2, which is a very neat drop-in replacement for the kit's empty space; the one significant omission from the kit.

Scale Aircraft Conversions

The aftermarket company, Scale Aircraft Conversions, produce white-metal nose and main undercarriage sets for 1/144 scale and 1/72 scale V-Bombers which can be recommended.

Set 14410 is for the 1/144 scale landing gear for the Great Wall Hobby Vulcan kit and the company have informed us that they will produce an undercarriage set for the Great Wall Hobby 1/144 scale Victor.

In 1/72 scale, Scale Aircraft Conversions produce undercarriage sets for the Airfix Vulcan (72020) and Valiant (72035). There are no SAC undercarriage sets for the 1/72 Matchbox/Revell Victor as the company remain in the hope that Airfix or another manufacturer will release a new tooling in the not too distant future...

It should be noted that Scale Aircraft Conversions clean up all of their white metal

sets, removing any seam lines, ejector pin marks and the like. Regarding their sets in general, if the sets do not contain corrective alterations they are referred to as 'replacement' sets; if corrections have been made to the original plastic kit parts they are referred to as 'improved' sets.

Alley Cat

Vickers Valiant 1/72 scale resin and brass etched update kit Maxi Set.

This set includes all three of Alley Cat's previous separate sets for the Valiant covering the engine air intakes and jet pipes, tailplane and wingtips and the mainwheel wells and doors into one single boxed set.

Freightdog Models

Freightdog Models provide two tail pipe sets for use with the Airfix 1/72 scale Vulcan B.2 kit.

Set 7201 comprises resin tailpipes as fitted to the Olympus 201 powered Avro Vulcan B.2 (early) variants and Set 7202 comprises resin tailpipes as fitted to the later Olympus 202 powered Avro Vulcan variants, including tanker, maritime radar reconnaissance (MRR) and also applicable to the last airworthy example XH558.